ROUTLEDGE LIBRARY EDITIONS:
SCOTLAND

Volume 28

THE POLITICS OF ENVIRONMENT

THE POLITICS OF ENVIRONMENT
A Guide to Scottish Thought and Action

MALCOLM SLESSER

LONDON AND NEW YORK

First published in 1972 by George Allen & Unwin Ltd

This edition first published in 2022
by Routledge
2 Park Square, Milton Park, Abingdon, Oxon OX14 4RN

and by Routledge
605 Third Avenue, New York, NY 10158

Routledge is an imprint of the Taylor & Francis Group, an informa business

© 1972 Malcolm Slesser

All rights reserved. No part of this book may be reprinted or reproduced or utilised in any form or by any electronic, mechanical, or other means, now known or hereafter invented, including photocopying and recording, or in any information storage or retrieval system, without permission in writing from the publishers.

Trademark notice: Product or corporate names may be trademarks or registered trademarks, and are used only for identification and explanation without intent to infringe.

British Library Cataloguing in Publication Data
A catalogue record for this book is available from the British Library

ISBN: 978-1-03-206184-9 (Set)
ISBN: 978-1-00-321338-3 (Set) (ebk)
ISBN: 978-1-03-207650-8 (Volume 28) (hbk)
ISBN: 978-1-03-207682-9 (Volume 28) (pbk)
ISBN: 978-1-00-320825-9 (Volume 28) (ebk)

DOI: 10.4324/9781003208259

Publisher's Note
The publisher has gone to great lengths to ensure the quality of this reprint but points out that some imperfections in the original copies may be apparent.

Disclaimer
The publisher has made every effort to trace copyright holders and would welcome correspondence from those they have been unable to trace.

THE POLITICS OF ENVIRONMENT

Including

A Guide to Scottish Thought and Action

by Malcolm Slesser

London · George Allen & Unwin Ltd
Ruskin House   Museum Street

First published in 1972

This book is copyright under the Berne Convention. All rights are reserved. Apart from any fair dealing for the purpose of private study, research, criticism or review, as permitted under the Copyright Act, 1956, no part of this publication may be reproduced, stored in a retrieval system, or transmitted, in any form or by any means, electronic, electrical, chemical, mechanical, optical, photocopying, recording or otherwise, without the prior permission of the copyright owner. Enquiries should be addressed to the publishers.

© Malcolm Slesser 1972

ISBN 0 04 320074 5 Cased
     0 04 320075 3 Paper

This book is dedicated to My Neighbour

Printed in England
in 11 point Baskerville type
by Clarke, Doble & Brendon Ltd
Plymouth

CONTENTS

Introduction		*page* 9
Part I	Nations at the Crossroads	13
Part II	The Scotland of Tomorrow	25
Part III	1. Scotland Today	49
	2. Resist and Achieve	100
Part IV	1. Constraints on Expansion	119
	2. The Balanced Communities	141
	3. Resource Management	154
	4. The New Politics	164
References		175

INTRODUCTION

I, Malcolm Slesser, am really only the editor of this book, presenting here some of the ideas being worked over by a growing number of those responsible for advances in science, technology and commerce in Scotland and elsewhere. These people, themselves specialised products of modern society, are not only aware of the malaise now obviously affecting this society, but by the very nature of their work have been forced to recognise its cause. The rapid growth of technology, so beneficial to man, has precipitated a crisis in the evolution of his society and, as in previous evolutionary crises of other living beings, an initially useful feature has galloped away out of control; our crippling dinosaur armour, our self-destroying sabre-tooth, is industrialisation. This excellent servant is a deadly master. Encouraged by the more ambitious practitioners of the as-yet naive 'science' of economics, this expanding industrialisation is today destroying the land and its communities, and tomorrow there must be heavy retribution.

The protests of the conservationists, the apparent unreason of affluent strikers, the violence of students, the attempts to escape through drugs into fantasy, are all symptoms of our growing unease: for it is obviously a great strain to pretend that we can continue to cram an infinitely expanding economy into a finite land – something must burst, and the greater the pressure the bigger the bang. But how can we stop it? Should we ourselves even bother? The ever-stretching affluence might just last out our time – our children can inherit the ruins. . . .

For certainly our time is short. Affluence no longer decides our consumption, but is itself solely and precariously dependent on the accelerating increase of this consumption. Hence the desperate search for wider boundaries, like that belt without a buckle, the European Common Market. But these makeshifts are of little use, as is increasingly evident; based on the sickness they try to cure, they merely exacerbate the situation.

A return to a balanced economy, where technological progress plays its correct part subservient to the individual, is obviously necessary. But such a return is extremely difficult for communities already violently pledged to imbalance by over-industrialisation and sometimes by over-population as well. Such communities expand forcefully, with the most plausible excuses, into their more favourably situated neighbours – usually after weakening any economic or cultural resistance. The USA has Canada, Mexico and the rest. The USSR has its vast eastern territories and satellites. England, more constricted and more desperately committed, has now only Scotland and what charity Europe may allow her.

The less committed nations have a chance to readjust with less upheaval, provided they can ward off the surrounding pressures. This book examines particularly the position of Scotland. It offers merely an outline, but its deliberate simplicity should not be taken as the last word; more detailed treatment of the subjects has been under way for some time. We have here tried only to put across the message.

For a message is badly needed. The planned flooding of Scotland with English workers and industries cannot possibly cure England's imbalance; but, to gratify the temporary indulgence of her addicted neighbour, Scotland as a community will be destroyed permanently and the regeneration of England made more difficult – for there would not be available the example of a restored and progressive society over the Border.

This book is in four parts. Part I examines the predicament of today's advanced nations from first principles, treating a self-governing community in the only practical context possible – its territorial and marine resources, the number of its inhabitants, and the cultural cement that has held the structure together through the past and will strengthen it for the future. Part II, taking Scotland as one such community, sketches the programme of its repair to a balanced and progressive nation, able to develop its own pattern of democracy in a world certainly needing as many such examples as possible. Part III illustrates how the land, the spirit and the future of the Scottish people are disappearing under sterile colonisation by an outdated 'British' economy, and describes how each member of the Scottish community can play his or her part in defeating this planned eradication, and so save the promise of the future for our children.

These three parts are deliberately brief. They are to be read as a handbook to further thought and action, as a framework round which discussion groups can build a more comprehensive structure. The authors have met so many who hold these views and have so signally failed to find them printed anywhere in relation to Scotland, that they judge such an abbreviated treatment to be necessary at this time.

The fourth part gives a wider background to the first three. It illustrates how leading scientists and economists confirm the dangers of runaway expansion of industrial aggression in a world growing smaller and indicates the disasters which await continued biological imbalance in a world becoming more crowded; in this context, it examines further applications of technology for the more satisfying Scotland of the future. It may be read as complementary or supplementary to the first three parts, or simply as a beneficial and instructive irritant in its own right.

Finally, we believe there is a need in this world to explore new forms of politics and government, based on the future circumstance of Man. We believe this will result, not in ever larger economic groupings, but in an emphasis of the role of national communities. Present-day Scotland is an excellent model to work on. Each inhabitant of that or any similar community can play an active part in bringing about this most necessary change. The newness of this book lies in this dovetailing of new ideas with an old struggle, rational observation with political evolution, theory with simple practice.

PART I

Nations at the Crossroads

Even as recently as 1967 it was possible for a major work to be written about the state of the world some thirty years hence, considering only economic factors. This book, *The Year 2000*, by Herman Kahn, prophesied dazzlingly high figures for the national wealth of many of today's industrialised countries. These figures took no account of the impact of unthinking industrialisation on Man and his environment – or that a reaction, often religious in its intensity, might interfere with the projected course of economic growth. This reaction is a natural one, being biologically necessary. Before it is too late for all of us, it is necessary to stop and consider the consequences of the obsessive industrialisation which exists today in many of the world's communities and which may well exist tomorrow everywhere.

As we do this we find ourselves inevitably asking for whose benefit a community is to be run. While obviously we must progress towards some form of world government, it is equally obvious that no centralised government could meet the wishes of the vastly differing communities which will continue to inhabit the earth, each of which has its own satisfying pattern of life. Within the world framework there must operate a number of reasonably self-governing units. We are at once led, therefore, into requiring a working definition of such a community – a nation. And it need only be a working definition – no absolute one is called for. As we consider the consequences of this thoughtless industrialisation, we will find ourselves arriving at such a working definition.

The main problems of the future therefore appear to be: firstly, the size of these self-governing units in a world state, and secondly, the day-to-day running of these units so that each person in them enjoys the greatest choice of physical and mental activity permitted by the elected administrators of his unit community. These local governments must provide basic aid against a member's misfortunes, such as sickness and resultant poverty, and against the destruction of his physical and mental heritage – his satisfaction in his land and its culture – by acts of aggression. Such aggression – waged internally or without – formerly was military, but is now increasingly and more effectively, economic.

In the past the determinants of the unit communities of the world have been basically two :

    (i)   the resources of the land occupied
    (ii) the number of people in that land.

But these are lifeless and statistical parameters; the stability of such communities, their inspiration to progress and the joy of living in them has been due to a third factor – the unique human awareness of time. By this great gift the past history of the community has been cherished, creating a common bond for the present; and this bond has been immeasurably strengthened by the promise of a future. In this way the community has identified itself and become aware of the possibilities of development; it has stabilised itself and thus been selected for successful evolution.

Communities have trodden along this path of time, yarning of scenes past, planning the ways ahead. Their cultures – their song and story, have blossomed along the route, with squabbling, theft and interchange. Many communities have fallen by the wayside, some victims of disease, flood or other unavoidable catastrophe, others overtaken by their more efficient neighbours.

Efficiency for these human communities, as for all living systems from single cells upwards, consisted in coming to terms with their environment, extracting from it their energy, their food and materials, mastering it sufficiently to serve their progress, yet not exploiting it so that it ran dry on them at their sudden hour of greatest need. History records the unsuccessful, the over-conservative stagnators who never tried and the brash adventurers who tried too hard, as fossils tell us of the earlier biological failures.

The most powerful material factor in the successful evolution of a community was the development of tools, fashioned with ever-increasing ingenuity to win more wealth from its own land or from the hoards of other communities; an industrial expansion by peace or by war.

Until a century or so ago this evolution of communities proceeded relatively slowly. The three determinants of land, people and culture interacted to preserve viability in the successful ones. Of course the stagnators were still being enslaved, the sated empires still toppled; but the world was very large and it had plenty of time. The mistakes of Egypt, Carthage or Rome affected only a small area.

But then the scene changed. Human social evolution began to accelerate rapidly. Such evolutionary spurts have happened to many species in the past, incurring often their salvation or their downfall. Science, with its accurate observation, opened up the secrets of the environment, technology spread this knowledge and, multiplying its acquisition at undreamt-of speed, used it to wrest further and further wealth from the land. The world became smaller and its dominant societies found one aspect of their culture, that of commercial industrialisation, growing to grotesque proportions, taking over as the sole determinant of a community. However beneficial any such aspect may be, its monstrous and unpremeditated growth at the expense of all others can only develop from the bizarre to the pathological; the pathological state is entered when the growth becomes malignant, a cancer which can no longer stop proliferating, which swells at the expense of all the other resources of a living community, and which will inevitably destroy that community.

Most of the large industrialised nations of the world are

entering this irreversible stage, and are no longer balanced evolving communities of the type we have described. We can recognise the fatal symptoms very easily. In face of such malignant industrialisation, the other determinants go down. The natural resources of a land are considered irrelevant, as are the number of people pushed into it. And, of course, the communal spirit of those inhabitants is necessarily eroded by this malignancy, and is actively attacked should they dare to oppose it. We are tragically familiar with the parrot cries of 'Bring in the jobs', 'Mobility of labour', the jeers at 'idealistic scientists' (who should stay at their benches), at 'romantic traditionalists', 'middle-class conservationists' and the like.

The tragedy, of course, lies in the naivety of those supporting the growth of imbalance. Some admittedly are merely cynical, content to be going back on human progress, to be replacing modern man by a collection of industrial helots, slaves once again, dependent on factory planners for their land, their culture, their free gift of 'instant community'. But most industrialists and their advisers are simply ignorant of the astonishing archaism of their approach. They cannot see that they break the elementary rules of the science they so frequently invoke, that they betray the technology they hire.

Science is largely accurate observation coupled with commonsense. The man in the street is, because of this, scientist enough to sense that something is going badly wrong. What is going wrong? What do we see around us?

We see a loss of communal spirit, of 'morals', of respect for any standard of ethics in the members of the present self-governing units.

We see, as well as the waste of these psychological resources, a waste of the material resources in the land occupied by those communities.

We see a growing backlash of resentment against such losses and ultimately against the whole politico-economic system encouraging them.

We see excuses offered, apologetically, that it is impossible to halt – in these finite, limited areas – what has now to be considered infinite, unlimited industrial expansion.

We see successive governments attempting to stabilise society by using this runaway accessory of a balanced development – industrialisation – as a final determinant of the self-governing communities, one which dictates the resources of the land and the number of people in it. And which therefore dictates the boundaries of these communities and their day-to-day running, the work and leisure and education of their inhabitants, and which no longer nourishes an individual person on his community and land, or on its history and its future, nor on his own free ideas, but which subordinates him to a confessedly unpredictable and as yet incomprehensible process of economics – a 'discipline' whose development is still at the leech, purge and incantation stage of mediaeval medicine and as disarmingly arrogant. The wiser economists will freely admit this; they are among those most disturbed by an 'expanding economy', as will be plain from the extracts quoted in Part IV.

Until recently this unease, felt by both the man in the street and the economist, has been bribed away by the attractive flashing of cash, the carrot of rising wages dangled on the string of inflation.

The sorry comedy of this diversion is now realised by many people, and is guessed by most. Indeed, the results can no longer be hidden, even by the most careful erection of outmoded political screens – such as the old barriers of 'class', or the newer ones of 'age'.

Those accustomed to enjoying their heritages of traditions and landscape, now burst out in demonstration against Concorde, Cublington, Hunterston, Teesdale and the London Box. Those brought up to regard skilled labour as no degrading thing and now, thanks to technology, enjoying at last comparative affluence, indulge in otherwise meaningless strikes in order to preserve at least the illusion that they still possess independence and self-respect. Those educated to the ideals of the old communities and finding no ideals left, nor any communities left to use them for, nor any freedom, even in a vacuum, to express them – these young people naturally escape to drugs or violence, whether in the USA, in Europe or in the USSR.

At first all this resentment had been held in check by the knowledge that science and technology have brought release from desperate poverty, disease and violence; that they were, in fact, useful tools of the community. Then the enslavement of individuals and destruction of communities by over-industrialisation burst the self-restraint, and not only the industrialists but the scientists and technologists came under violent attack.

Now – and here is the main message of this book – it is dawning slowly that the use of scientific knowledge and technological skill for the conquest of poverty and disease does *not* of necessity mean enslavement to expansionist economics; that this last is an aberration itself bringing inevitable misery, violence and poverty, whereas science and its associated skills remain vital tools, but tools only, of a community: and communities need not – *must not* – be overwhelmed by a vast unnecessary mistake. In fact, science and technology provide the only means of removing the mistake and repairing the damage done. This realisation is strengthening the protests by focusing them against the mistake, not against the scientific evolution of the communities. Therein lies the great, the only, hope for the future, preserving at once our amenities of civilisation and landscape, and our self-respect, and once again allowing valid ideals for the sorely-tried youth of a community.

Of course, not only the governed are disturbed by the gallop to disaster. Governments themselves are making desperate attempts to draw boundaries and lines to contain, and somehow rationalise, the economic expansion of certain communal groups. All these expediencies are doomed, for the groupings are based on superficialities; on the temporary picture of the ever-changing economic pattern of the past fifty years. Economics is still the sole determinant; economics, the most unstable of factors, whose uncontrollable expansion is itself the reason for these desperate measures, is to override all the other factors of land, peoples and cultures that have at last been worked in, after thousands of years, to some sort of harmony. Naturally, the attempt will not succeed. The present problems will get worse. The parcel is too crudely packed. As the EEC string bites into the national nerves of, for example, that remarkable community, England, the protests are indeed loud.

What remedies are possible? How can this career down an evolutionary blind-alley be halted and the committed communities of the world resume their correct course?

The large 'advanced' nations are already so far committed that their size makes it most difficult for them to slow down and get back on the path. In these nations unrest must be expected to increase for a long time yet. Small advanced nations are usually, because of their better biological and economic balance, not so far committed, but are imminently threatened, for they can be taken over as temporary shots-in-the-arm for their bigger failing neighbours – a tragically useless sacrifice.

For a sane assessment of the possibilities of restoration, let us return to the realities of our determining factors for viable communities, viable in their *total* environment in all four dimensions, not just 'economically' viable. Let us consider communities that are at present, or have been in the recent past, accepted as 'nations', and whose members possess a communal belief in what we can only call from long human experience a 'national' spirit; and who are potentially able, in emergency, to be fed, by modern biologically acceptable methods, from the land they inhabit.

Let us start with those small nations not yet committed too far along the Gadarene slope. Let them be allowed to make themselves practical examples of a realignment to balanced development, in cooperation, and by the *humbly* scientific use of their physical and mental resources. And their larger and more committed neighbours could scarcely fail to benefit. We need as many pilot examples of democratic balanced communities as possible; a *modern* form of government scarcely exists as yet, one utilising the immense power of technology to serve the community, to ensure satisfying employment, and leisure and to nourish its hopes for the future.

If we look around we see several nations existing as almost immediately salvable communities; Norway, Sweden and Finland are three examples. Other communities are salvable but are threatened by takeover, some already half-submerged; Canada and similar American satellites of the USA, the Russian economic dependencies, and Scotland.

Scotland is a good example in the island of Britain, probably the only practically salvable community there, being blessed at present with a reasonable ratio of population to area and with resources which are now useable by modern techniques and capable of being protected by a communal spirit. Wales is grievously interpenetrated, held together as a community largely by a language less and less withstanding cultural aggression. England, with one of the worst ratios of population to area in the world, appears, increasingly and almost by choice, committed to overpopulation, over-industrialisation and racial problems, and has yet to face readjustment to the eclipse of her recent moral and physical overlordship of the world; she is a gallant but misguided community and a most dangerous neighbour.

Paradoxically – but on reflection, of course, obviously – the problem appears less evident in salvable countries like Scotland than in those as deeply committed as England. And naturally the London Government, anxious for the brief respite of an overspill to the north, and jealous of the brighter future available to their Union partner, encourages this blindness. Many Scots, with such assistance, welcome a future of 'British' industrialisation, as a relief from their present neglect. But it is not for nothing that a remarkable number of highly skilled immigrants from south of the Border are actively opposed to a

'British' development of Scotland. They know why they left England – they know what it's going to be like.

Obviously it is urgent to begin the work of salvage *now* in *any* community. 'Now' is the gathering twilight of today's pattern of economy. Its ending is visible to some, foreseen by many, but to others as yet unimaginable. These last are joined in their happy *laissez-faire* by the apathetic and the cynical, victims of the unrooting process, of the propaganda against national and community spirit. They are joined also, and this time actively, by those who see clearly the forthcoming general disillusionment and hope to exploit it with the help of the bewildered unemployed: these are the contemporary revolutionaries, Fascist cliques of the future, that London, Washington – and Moscow – fear so much and yet so indirectly encourage.

No political party will gain from unchecked industrial expansion. The Right must lose its traditions, its leisured privileges and be blamed in the crash for its selfish exploitation. The Left will also lose its traditions, must sacrifice the independence of thought and choice of the workers, must condemn them to stultifying tasks under hierarchies of managers; and will, in its turn, be condemned for exploiting society by class strife. The Communists, who might hope to cash in on these troubles, will, if they survive their own or their opponent's police squads, be forced to restore some 'nationality' to their totalitarian states: just as the USSR became Mother Russia in the war and has since had to fight continually with 'national' ideals from Poland, Hungary, Czechoslovakia, Yugoslavia, Roumania and the rest, and as China has had to argue with Cuba and Viet-Nam. A community's sense of identity is not eliminated along with the bourgeoisie. Communism, whose great merit was to open our eyes to the dynamic evolution of communities, has remained mesmerised by one man's fixation on a single stage of a nineteenth-century industrial aberration; the evolution he recognised has left Marx far behind in his grave, but his followers still kneel there, as outdated as the even less perceptive politicians to their right.

For modern technology need no longer be the servant of economics. It is now able to halt expansion-for-expansion's-sake without entailing unemployment and recession. There is now no excuse whatever for trying to impose this servant 'economics' willynilly over the more biological, nourishing, attributes of human communities, trying to impose a theory however academically 'scientific' over the boundaries already determined and accepted by that much more practical experimenter, Human Evolution. We must be humble and observe. We must allow for those unacademic terms 'community spirit', 'national morale', at all times, not just in the practical emergencies of war. If we ignore them in our economic aggression we shall face the same consequences as if we ignored them during military aggression. Russia in Yugoslavia and Finland, the Reich in Europe, Westminster in its recent Empire, the Pentagon in Viet-Nam, the French at Diem Bien Phu, all acknowledged, after bitter lessons, the impotence of modern military aggression. And our evidence indicates that economic aggression will meet a similar impasse, with anarchy and totalitarian reaction making life unpleasant until the forces of Man's spiritual resistance, gathering underground, can once more turn the invader out. And all this tiresome struggle, like war, is so unnecessary.

It is surely common sense to use science and technology, not to force an antiquated economic theory to work against all the biological facts, but to make them work *with* the existing communities and to use their almost unlimited skills to produce a civilised, stable, fully-employed community in a satisfying land. It is extremely foolish not to try and do this as soon as possible in those communities – such as Scotland – that are, as we have seen, not yet very far committed to the mistake, and so have fewer readjustments to make.

But politicians are bound to the short-term view; and they are even more naive in science than they are in economics, though with less excuse. Understandably, they push on with failing hope, like Napoleon, or Hitler, to Moscow; they dare not think about a return. Now is the time, therefore, to check them. They cannot do it themselves, they cannot resist those twin sirens, the CBI and Transport House, and no Ulysses being among them, they have not insight enough to ask to be tied for a while to the mast.

We can remedy this more easily in Scotland, where, thanks to Westminster, we have no contemporary politicians worth worrying about; and we do possess a people able to elect a modern system of government, a people with commonsense, a living communal pride and tradition and a host of skills, manual, technical and scientific. We are angry and frustrated, as the great petulant vote for the SNP, that well-meaning old lady, indicated. But many of us cannot see how clear the way ahead really is. Others feel reluctant at 'leaving England in the lurch'.

It is one purpose of this book to go a little way towards showing what we can do in Scotland and that, so far from abandoning England, we can teach her, from our more favourable position, to correct her growing errors. Along with the other less-committed advanced nations, we have the chance to opt out, safeguarded by today's technology (as yet scarcely used for the *individual's* convenience), and create a satisfactory structure for our modern community.

That great petulant vote for the SNP had a positive element; yet it was a gesture of sentiment rather than of inescapable intellectual conviction – among many of the candidates as well as their electors. But an approach along the lines indicated in this book does lead us inescapably to the necessity of self-government for Scotland, to the means of establishing it and of ensuring its success. Sentiment had, as usual, its heart in the right place; we now see how we can use our heads.

The matter is urgent, for the Westminster government is at present taking active measures to destroy for ever any chance of our balanced development, and, as balanced development is a *sine qua non* for a self-governing unit in the modern world, such action can only destroy any chance of a Scottish nation. And by planned overspill of industry and workers from the south (surely this 'mobility of labour' is another Ultimate Solution) to lure away our most talented people and flood us with unskilled strangers, and so remove any possibility of organised community opposition to final exploitation. There will then be no Scotland; another community will have been destroyed. The world, we think, will be the poorer; we Scots have given it memorable examples of the free human spirit, our independence and our generosity in the face of brutality – think only of Galgacus, Wallace, the Arbroath Declaration, even of the miserable Highlanders after Culloden; our wit and learning – as of Edinburgh in her Golden and Silver Ages; our matchless song and story. We have international respect for all this and more, and the world would be sorry to see it go, for we have much to exchange with each other. But we shall not let it go. For if it went, it would not be a dream that had vanished, nor an 'auld sang', but the very future, now so bright with unsuspected promise, of our own children.

PART II

The Scotland of Tomorrow

Let us look at Figure 1. It illustrates one way of setting out the inherited resources of a community, and their interrelationship.

The resources are the given data; they determine the viability of any community. They are not remote statistics in some file. They decide to a very great extent whether we shall enjoy our life here on this part of the earth, whether we shall enjoy our jobs, our cities, our countryside, our sense of living together, whether we shall be satisfied we are leaving something of value to our children.

Yet, for Scotland, few official surveys have yet been made on any resource in any of the areas we have classified as territorial, social and psychological, virtually none of the resources of any one of these separate areas, and no study at all on their interrelationship. As all these resources mutually sustain each other and their strength is only fully appreciated in their harmony, then no wonder that no idea whatever of the potentialities of Scotland has been available to the man in the street.

He may have glimpsed a few of the potentialities, camping among the mountains, visiting a new hydroelectric dam, hearing a half-understood song, listening to the discussions of experts, reading some obscure and learned author, or arguing one memorable night in a public bar. Even from the sporadic literature of Home Rule political groups. But never from the Westminster parties, from the mass media, from his state education. Such a conspiracy of silence is of course only partly deliberate. Much of it derives from the current obsession with 'British' economic growth outlined in Part I. But the rest stems from the desire of the central government to keep these resources hidden, to deny their existence or to confuse them

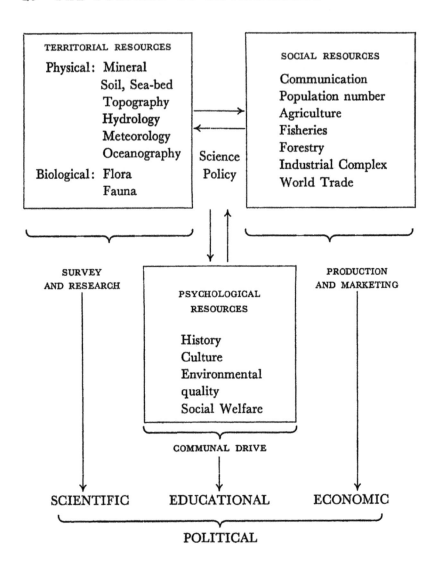

Fig. 1 The inter-relationship of inherited resources and their political harnessing for the future

with irrelevancies and 'humorous' caricatures; and this dishonest policy has to increase now that modern techniques are revealing more and more of Scotland's true resources. But we will deal with that propaganda, and the way to defeat it in Part III.

Here we will, very briefly, go over Figure 1, so that at least an idea of the wholeness of our inheritance can be gained; each individual may then explore his or her favourite corner of the framework and see how his work or pastime fits into the life of his community and enriches it – or could enrich it, and enrich himself, if Scotland were considered as a modern nation, a self-governing unit in the gradually evolving world state.

From Figure 1 we see three boxes, three areas of resource, three convenient classifications: territorial resources, social resources and psychological resources. Let us accept this and succeeding terminology for the purposes of the argument. Doubtless modifications can be made, probably several may have to be made; but let us not play the enemy's game at this point. If we are able to climb a tower and get any view at all, we should be grateful. We can always polish the windows later. We inherit these resources from our parents and the community. They are inter-dependent on each other, and any change in one affects all. Their mere existence does not predetermine the development of Scotland. Only when an informed government has instituted a co-ordinating system along the arrows of Figure 1 will the community be able to choose its path intelligently.

Let us firstly consider the contents of each box in turn, secondly the interrelationship of the boxes, and finally how the three areas of resource can be harnessed to ensure our satisfaction and the political viability of Scotland in tomorrow's world.

Territorial resources
These are the physical baselines from which the other resources operate. They cannot be enlarged, but *can* be explored

thoroughly and the results analysed by computerised techniques which at least enable reasonably systematic classification and retrieval of information. From this knowledge the government could plan conservation and harvesting measures to yield the most efficient social return. An environmental Information Centre would act as a data bank, to be drawn on by all enquirers; it would also contain similar internationally held data from abroad.

Presumably the extent of Scotland as known today will not alter; the days of territorial aggression are over. We shall either continue to possess our 30,400 or so square miles, or have nothing – as we saw in Part I.

This area is a handsome inheritance, sufficient to feed its present population in an emergency and afford them recreation in a topography as beautiful and varied as anywhere in the world; a terrain not subject to extremes of climate, abundant in water and coastal resources and possessing some of the finest harbours in Europe; a land that only needs informed and sympathetic attention to bear great fruits.

Its mineral resources have been examined over a long period; we can look forward to improved methods of extraction and the increasing use of others than the familiar iron, coal and quarry-stone; there are many intriguing deposits in the Highlands. We are fortunate in having already at Aberdeen one of the foremost Soil Research Institutes in Europe, ideally equipped to give us a comprehensive survey of our soils; from such information we would, if considered desirable, improve areas naturally infertile or rendered so by past opportunist farming.

Topography is at present well served by maps, but the practical applications require more detailed work and, as in all these instances, need to be considered in the light of the *needs* of Scotland. The lack of such a focus on Scotland has resulted in the present neglect or squandering of our resources. This is evident even in the meteorological and hydrological fields. We know the fate of fog-free Prestwick, considered too provincial as an international air junction. Again, the Western Isles and parts of the Highland seaboard are as mild as Cornwall and the Scillies, and like them would provide early potatoes and bulbs, once transport were organised for Scotland's benefit; without such a focus, we get the Highlands and Islands Development Board, able only to nibble at the prospect. On the mainland, how are 'rain shadows' in the lee of the mountains – and their converse areas on the other side – exploited? Where can water resources be tapped – not only from wasteful and now largely-harnessed surface reservoirs – but by drilling to an almost (for Scotland's purposes) limitless supply of ground water? In a self-governing Scotland, there will be no flooding of valleys, as in Wales, to indulge the thirst of runaway English industrial adventures.

Oceanography is a vital study. Scotland, with exceptionally large coastal resources, has scarcely begun to use them. Shore line studies, apart from providing an ecological basis for fish, lobster and kelp farming, are essential for evaluating the development of any marine-based processes: the present habit of dumping or pumping industrial effluent, radioactive or otherwise toxic, without intimate knowledge of tides and currents is deplorable. Oceanic resources are a matter for international arrangement, but to judge by present patterns are to be divided on a fairly equitable basis among the shoreline communities. Division of the North Sea for oil and other exploitation has already been made: the Scottish sector can easily be delineated. Scotland's deployment of this resource depends on her science policy (see later), but clearly the low-sulphur oil now being found in the Scottish sector represents a most attractive resource worth about £400M a year.

Biological resources, as defined here, are the indigenous flora and fauna of our land and coastal waters: the pines, birch, deer, grouse, salmon and other members of one of the most interesting wild eco-systems left in Europe. We are fortunate that information has for some time been collected by a devoted band of experts, inspired by the spirit of Fraser Darling and

Duncan Poore. Apart from its value in tourism, recreation and education, such knowledge is vital for any scientific appraisal of the result of projected new farming, horticultural, fishing or forestry techniques. To neglect or exploit those other living beings who share our heritage is to betray our human responsibility.

Social resources

The next box in Figure 1 summarises the utilisation of their territorial resources by the inhabitants of a community. Obviously the basic social parameter must be the number of these inhabitants. We can recognise a realistic upper limit of population: there is little practical fear, with modern technology, of a lower viable limit.

Any biologically balanced community must allow sufficient farming and fishing – or room for its expansion – to feed this population either in times of economic blackmail by neighbouring powers wrestling with their final problems, or in times of sudden world crisis from war, viral disease (now with high populations ever more likely), or geographical disaster. Modern methods of conservative farming make this, the most important policy of social insurance, possible in countries like Scotland. Agriculture should be spread among as many – not as *few* – inhabitants as possible by intensively worked cooperatives; this eases full-employment problems, when all other industries are geared to occupy the remainder of the labour force and to provide material for internal consumption and external exchange.

Some 85 per cent of Scotland, sixteen million acres, is classified as agricultural, of which 3,262,000 acres are arable and another million grass. According to Dudley Stamp's *Land Use Statistics of the Countries of Europe*, Scotland has at the moment some five million 'adjusted' farm acres. Since one 'adjusted' acre is sufficient to support one human being, the community even at its present level of agricultural development can support its five million people. It is thus a biologically balanced community, which is why we can talk about its future.

Although now extractive pastoralism has been competently developed on the low ground and recently upon the high marginal farms, repeated burning and former episodes of 'over-industrialisation', such as the early nineteenth-century sheep boom, have devastated large areas of the Highlands, so that they are now 'distressing to anyone with some appreciation of ecological principles'.[1] There is great potential awaiting recovery in these areas. Scotland can therefore assimilate some growth of population, which gives it time to gain stability. At what level stability may be achieved simply depends on the biological resources. Iceland would perish as a community if its population rose to one million; it plans therefore to allow a gradual increase to half a million, beyond which point it cannot expand in number without biological imbalance and therefore eventual destruction as a community.

This problem of course faces all the modern world, and we will discuss it in Part IV. As regards Scotland, the level is not known with any high degree of certainty until we have our land-use survey. However, given our agricultural potential, which we have seen could be higher than it is at present, Scotland could have six to seven million inhabitants. At that point population density would have reached 233 per square mile; it was 171 in 1971. At present it seems sensible that this natural population growth should expand beyond the already dense midland belt.

It would therefore be the fundamental business of the government to see that this population was adequately supplied with varied foodstuffs, through farming, fishing and export exchange, and that the internal food production was maintained at a high enough level to cover at least three-quarters of the need, with fallow areas to call upon in emergency. Alongside must run a well-developed food-processing industry; Scotland already has several research institutes devoted to such processing, and the community is capable of being among the world leaders in this field. Fishing, whether as fish farming, coastal fishing or deep-sea trawling, needs no further comment here; like Iceland, we must strictly protect our inshore fish nurseries against aggression, for our future harvests depend on such conservation.

All these forms of activity provide adventure and self-respect for many people to whom indoor jobs would prove intolerable, and such people are very necessary indeed for an alert 'psychologically-balanced' community. The driving of these folk into unimaginative factory halls – or more usually, to emigration – is not the least evil of today's industrial folly.

Forestry is another open-air task, and Scotland has peculiar advantages for it, especially for raising the highly productive north-western American coastal trees. More forests are being planted in Scotland, but as another example of economics dictating the needs of a community instead of the other way about, we see our lauded forester training schools (as at Faskally), being sold off and skilled men asked to retire early, their places being taken by temporary gangs of mobile labourers. Where is the life, the self-respect, the employment that forestry was to bring to the Highlands, as Mr Edlin assured us in all those publications? Such things have not turned out to be 'economical'. But as we showed in Part I, no living community can be 'economical' under the present 'British' outlook. In an independent Scotland the correct priorities can be restored and expanding forestry will fill again the empty places.

Even on a world-wide basis demand for timber consumption is greater than its production. Timber is a remarkable material. Structurally, as beam, plank, laminate or compressed board it possesses a unique combination of strength, lightness and ease of handling. It makes insulation, paper and wrapping; its bark is a useful fertiliser and mulch. Nowadays every scrap of it can be utilised, and modern methods can render it virtually rot and insect proof, and remarkably fire-resistant. If we had invented it, what a revolution! Moreover, its very production improves the landscape. Fortunately the present monocultures would give way under the thoughtful management to more biologically-valid – and so more recreationally-enjoyable plantations; if not to that ideal mixed sustention forest of Mark Anderson, then at least to more intelligently varied groups less susceptible than those monocultures (imbalance again!) open to catastrophic destruction by insects, fungi, fire or storm. We may thankfully bid farewell to those 'tough' foresters who, by aerial spraying of toxins, contrive, at the expense of our environment and health, to keep these monocultures going for the sake of some imaginary economic situation forty years ahead. Our forestry will not be abstract theory, but integrated with the Scottish community.

Only then, with the residual, but still large labour force, would one turn to the other industries, to services, research and development. To our present back-to-front vision, this may seem retrograde. But it is not. The primary producers of food are essential to a balanced community; moreover in Scotland, even on present reckoning, their actual productivity is comparable with those engaged in secondary industries.

Some disgruntled people have suggested a moratorium on technological development. This is not only unlikely, but impossible. Technology is part of culture. Man needs an element of conflict to raise himself above the vegetable. Whether individually we like it or not, collectively we must strive. Here in our shrunken globe, with limited resources and space, we have only one unlimited ingredient, and that is the power of the human mind to develop new ideas. Since it is obvious that growth in quantity cannot persist much longer, growth must take the form of growth in quality. Stagnation cannot be acceptable to a viable community. Its energy will take two directions.

One direction is that of leisure activities. These are very much dependent on space, and a modern society will try to command as much space as possible. In Scotland, again we are lucky. Merely by controlling this resource we will be able to pick and choose amongst those gifted people who want to belong to our country; we can even see this tendency in today's twilight by studying the phraseology of the advertisements for high executives put out by enterprising firms situated in Scotland (as California is – or was – the desirable location in the USA).

The second direction is in the development of industries that suit our resources. In this the first stage is the creation of new and profitable technologically-based industries. Even in today's amorphism Nigel Calder can report that 'Scotland, an underprivileged part of Britain, has by regional will-power built up the largest electronic complex in Europe'.[2] In an independent Scotland the income from these advances is needed to modernise the country. Development incentives, which in the past have only distorted the economy and favoured the established trend, must be directed towards stimulating pioneering talent in innovative industries. Of course, research and development are not enough; they must be accompanied by intelligent evaluation and support of the ideas and inventions appearing.

There is no truth in the dismal picture continually – and understandably – put out by the London government. Scotland's industrial future and export prospects are excellent. Because of her favourable biological balance her options are vastly greater than those of England, who, striving to keep her swollen population working and fed, must always be open to economic blackmail from abroad. Scotland, even in her present state, has been for years a successful exporter with no problems of trade deficit. Even under her present wasteful and ill-directed industrial system she managed to export beyond the UK in 1970 some £766 million, about £148 a head – a world record, though small good it did to her inhabitants, shackled as they were to one of the most mis-managed economies in Europe.

However, all activities require physical communications and these are poor in Scotland today. This inefficiency reflects only the incompetence general in communications administered by the central government, and is not due to Scotland's topography. One merely has to consider the difficulty of travel in England itself, how great a barrier the petty hump of the Pennines is to any road or rail journey east or west, how long the motorways remained unbuilt, how curiously ill-considered the rail service still continues to be.

Much is made of topography hampering potential transport in Scotland, with our mountains, firths, sea-lochs, islands and remote settlements. Naturally, physical problems of communication are different from those in England, which should in itself recommend their consideration by a separate authority. But difficulty of terrain or coast is incomparably less than that suffered by, say, Iceland or Norway. Iceland, with no more people than Dundee, scattered over an area larger than Ireland, has a steep coast of fjords and an interior shaken by earthquakes or eruptions and barred by floods and icecaps; Norway, with less people than Scotland extended over the distance of London to North Africa, is mostly glaciated mountains rifted by deep valleys, and her coastline, the most intricately-fjorded in the world, is studded along its 12,500 miles with innumerable inhabited islands. Yet both these independent nations possess excellent communications by air, sea or land, enabling remote places like Siglafjordur or Höfn to be within a short time of Reykjavik; while Hammerfest, Tromsö, Narvik and Bodö, far above the Arctic Circle, are typical thriving modern Norwegian towns with a standard of living laughably far above that of Skinflats or Cowdenbeath, let alone Kyle or Stornaway.

What causes this difference? Scotland's hills are small, crossed by low passes; she has few islands and fewer fjords; in the midland area, notably, there are plenty of good-surfaced roads. Yet communication is so difficult, and goods transport, especially in the north, is unpredictable and expensive enough to crush enterprise: such transport is needed before a remote settlement can become viable, yet it has to become viable before it can pay for the transport.

Obviously this, like other related problems, is due to misdirection, rather than lack, of resource. A Scottish government would, in order to utilise Scotland fully, necessarily provide the varied modern coastal transport, the village airstrips, the tunnels and roads (which, even if often unmetalled because of severe winters and inordinate length, are at least *there*) so evident in Iceland and Norway. The problem cannot be planned, let alone solved, by a centralised government controlled by a distant majority. Our present roads, significantly, are numbered, A.1. and the rest, in relation to their importance for London; railways are timetabled for the convenience of London. The map shows the insensitive distribution of both road and rail, draining out of the country down to the south. Their future in an expanding 'British' economy would be to remove our resources and pour in congestion. These systems are arteries of transport and nerves of communication; their re-design and integration are vital to the development and maintenance of the living community of Scotland. It is hardly likely with our small area and large resources that we cannot manage better than stretched-out Norway or tangled-up England. An accessible and economically healthy periphery is vital to independent Scotland; it must remain merely a vote-catching paragraph for a Westminster government which has increasingly more pressing distractions.

Psychological resources
These allow the balanced community to survive as a nation. They provide the flame without which other resources will never yield energy. In contemporary Scotland this flame has been beaten back shamefully, but still smoulders away beneath. The present waste of our psychological resources is deplorable; a tragedy for our community, a disgrace to contemporary government. It is small consolation that this government, stricken by the 'British' death wish, allows England's own flame to die, while considering ours still bright enough to be worth extinguishing.

The core of the matter is education; this is the transmission of knowledge within the community : horizontally from expert to layman, vertically from old to young – with plenty of necessary corrective feed-back in both cases.

This knowledge is not only of the technical kind we have examined in our other two boxes, but a knowledge of the human community itself. Only in this way can a community become aware of itself and *will* its own development to the future.

It should be needless to state that the study of history is vital in this self-identification. History should start with the local group: history of the local landscape, the first hut-circle builders, the megalithic astronomers who took observations on that very hill behind the town (you can check their sightings today, thanks to Dr Thom); the history of the first tribes, chiefs, the temporary raids of Romans, the Picts and their remarkable carved stones (you can take rubbings of them now, yourself), right through Scottish history to the – we hope by then, more satisfactory – present. And hand in hand with this local history, which expands to include all Scotland, an outline of the world's history, not just European, still less just English history. Today's children are robbed of any such community story – of their own community and of the world community, both of which are to be of so great an importance in the future.

Children in an Aberdeenshire country school are taught about King Alfred and the cakes; they cannot equate King Alfred and his Wessex cakes with the battles fought in their own fields and farmsteads, the battles of which their berry-field ballads and bairns-rhymes tell them; what was Harlaw, why was Airlie burnt? Rejecting Alfred's cakes, they reject all, and grow up conscious of being on the English educational periphery, no matter how subsequently successful. In almost every Lord Provost, in almost every Scottish MP, you will find this 'provincial cringe'. But more of that in Part III.

So the young should be shown this vital thread of continuity from their own hut circles to their own city, from the local battles of long ago to the coming unification of world civilisations. A glorious inspiring story, when they are part of it and their Scotland is part of it. And is such education not one with the biological balance we are insisting on in this book?

Exactly like this should be treated geography and culture (of language written and oral, of music and other arts), beginning with the local and recognisable, and relating it to the wider and more difficult.

Scotland has had an astonishingly varied history: its bravery and its treachery, together with the beauty and savagery of her scenery, fire her song and story. Her Gaelic inheritance – including the unique classical music of *piobaireachd* – is world renowned among the knowledgeable. But how many of her children are taught anything of them? How much do her adults, her SNP-voting enthusiasts, know of this enviable heritage? To pick random examples – the exquisite early Gaelic nature verses; the great sea poem of Alexander MacDonald; Dunbar and the makars; Urquhart; Scott, acknowledged for the wrong reasons, while his imaginative peer James Hogg, author of the finest poetic parody and one of the most remarkable novels of the nineteenth century, is presented as a pawky clown; Davy Hume, whose bland destruction of empiricism still smoulders beneath the philosophy of modern physical science; Fergusson, Cockburn and that brilliant political circle; the magnificent best of MacDiarmid; that architect Mackintosh who so nearly showed how to turn the Scottish vernacular into a modern idiom and so, by civilising the flinty Bauhaus, might have created what we all so desperately need, a fertile contemporary architecture. . . . Why go on? The wonder is that the flame of our psychological resources keeps alight at all under the damp blanket of the present educational system. There is some power in that flame, to keep going as it does on scraps of Rabbie Burns, snatches of Loch Lomond, bits of Royal Stewart tartan and irrepressible anecdotes like The Wee Magic Stane.

Culture, then, the commentary throughout the years of our

community on its history and its land, would be the central nourishing feature of the new Scottish education. Only within its context, and with its comments, can our technological advances achieve their full value or even get off the ground in the first place.

With such an education a community cannot help looking to the future: the thread of continuity leads there. It also cannot help developing a social responsibility. And this responsibility is of course helped by the small easily-comprehensible size of the community: and the administrators are readily accessible, readily changeable, not hidden away down south. The present remoteness of administration in today's large unbalanced states leaves the individual the impression that he counts for nothing. And where nothing is offered, all may be taken: 'Us' versus 'Them'. The new rootless generations in the super-states have to find substitute 'mother communities' in vague shifting associations, in the Universe itself – the latter notably unsympathetic to the ingenuous unfledged. No wonder they break down so readily to become drop-outs or commercial fodder.

Together with social responsibility in the community goes, of course, governmental care for the sick and old; most countries by now have caught up with the standards introduced by the small advanced nations of Scandinavia nearly fifty years ago, and one can scarcely expect Scotland, of all communities, not to be among the leaders in improving such a vital service.

Another important psychological resource, at present being belatedly recognised, is the quality of the environment. Scotland has inherited some of the finest natural scenery in the world, and also some of the worst industrial landscape. The first is an inspiration to be preserved with all power, the second an intoxicating challenge – like a dirty room to a scandalised housewife, or an upset work-bench to a determined craftsman. There is work enough for a hundred years, for artist, tradesman and technologist, in the physical repair of our stricken Midland

Belt – that grim reminder of the beginnings of our imbalance. The more recent ugliness of brash, unthought, contemporary development is less substantial and more readily made satisfying, but only if we take over our control very soon: otherwise we shall be engulfed, from Forth to Clyde, in the mortal predicament of Los Angeles.

There is no reason why every inhabitant of the new Scotland should, like his present-day Scandinavian counterpart, not only live in clean enjoyable towns, but also camp and wander in the mountains, sail on the lochs and seas. All this is his natural inheritance, and the only restraint would be to preserve this splendid opportunity for others after him. Ironically enough, it has been those stage villains, the great sporting landlords of the past – albeit unconsciously – who have kept us our landscape unspoiled. Their present-day successors, the impersonal London business syndicates and the commercial exploiters, armed with mass-persuasion techniques, are far more dangerous; they require strict control in a balanced community, and will certainly obtain it in a socially conscious one. Tourism, though a great industrial asset for foreign exchange, must like every industry in a modern nation be subservient to its community's enjoyment of the land; we will have first choice of our own Scotland.

The sum effect of the utilisation of these psychological resources is to nourish a healthy pride in one's community; not to the detriment of other communities – they are merely different, not worse. This clean breeze of satisfied nationalism blows all chips off the shoulder and promotes internationalism of self-respecting and mutually-respecting communities; it is implicit in the substitution of a Commonwealth (however reluctantly conceded and shakily delineated) for an Empire (however well administered). With the resources available to Scotland *made* available to her community, we need have no fear of our future among the nations of the world, nor of the value of our independent contribution to a future world state.

Inter-relating resources
One duty of a modern Scottish government is, as we have seen, to organise information on these inherited resources and cause the gaps to be filled. Another duty is to inter-relate the resources, to co-ordinate all data so that their significance to the community may be assessed. As we see from Figure 1, investigation of territorial resources can be classed as survey and research, and is of a predominantly scientific nature. What we have termed social resources can be gathered under production and marketing, and is of a predominantly technological nature, the study of finance and economy being properly undissociable from such technology in any modern community hoping to evolve in a balanced fashion. Utilisation of the psychological resources, so necessary for the life of the nation, implies an educational approach, and this is more difficult than the others, for the approach must not be dictatorial nor meddlesome, but should exalt objectivity and criticism; it must show why all our varied resources are valuable, and yet refrain from making too many value judgements as if *ex cathedra*. This would be easy, however, compared with the effort of achieving any personal self-respect or communal pride in the present state of Scottish cultural education, which increases in degradation from the primary stage through to the universities.

Harnessing the resources
Much of the technique here is implicit in the preceding. We can only briefly mention a few guiding lines; the details may be left to the electoral programme of the new Scottish government, based on the advice of those experts now engaged in the work. Part IV will develop certain aspects more fully.

One early task of the new government of Scotland must be to develop economic research. The great problem will not be to generate growth by volume, but how to obtain growth in quality; how to halt wasteful expansion without incurring stagnation; how to ensure change, the necessary flux in society, without making unnecessary growth its objective. Because Scotland, unlike, for example, England, has the space to grow a little, she has time to find the solution to this problem; she has probably some twenty years at the most. It should be enough – if we start very soon.

Economics is a study which attempts to understand the interrelationships of production and markets, resources and consumption. It is a highly complex subject, especially as – despite the economist's wishes – people do not behave like a series of identical molecules. Understandably, to render the problem manageable, economists have reduced human systems to simplified models. Then, by making certain assumptions (in fact, intelligent guesses), it is possible to offer predictions. From this procedure, techniques of great value have emerged, such as econometrics and input-output analysis. But they are still very unreliable. This is not to decry economists' attempts. The same techniques have long been used in physical science and are more rewarding there because, unlike human beings, a million molecules of a given species can be predicted to behave, overall, in a consistent way.

However, the greatest weakness of the current state of economic 'science' arises from a further, fatal simplification; it tries to grapple with the problems of modern society by treating it in isolation from the rest of the world. This is akin to assuming that man's environment is infinite in extent. But it is not. The earth is in fact reasonably near to being a closed system, much as a space capsule is. The only significant energy reaching it is the sun's rays. The rest is in internal dynamic balance. When the population density rises, the 'infinite' model no longer holds. Man is now too numerous for economic theory to be valid unless it takes equal account of the environment. So this most vital study can only progress when economists appreciate that, however unpredictable he may be, man is not even a law unto himself, a separately assessable part of the world: man is part of 'biology'. The world is a delicately balanced eco-system and man is part of it – and moreover he is threatening the stability of the whole structure, himself naturally included. Man has reached the level of population when, by the evidence of many eminent people who have made the problem their life study, he dare no longer grow indiscriminately. Our passion for uncontrolled expansion, our greed for raw materials, our unconcern for waste, are dangers to our survival. Athelston Spilbaus, President of the American Association for the Advancement of Science, pointed out: 'As the standard of living goes up, the amount of waste and consequent pollution must go up . . . the next industrial revolution must be a planned one . . . on the thesis that there is no such thing as waste . . . there must be a loop back from user to factory.'[3]

So our economic considerations must move into the wider field of ecology. This is the science of the ultimate economy, the economy of nature. On his own, the economist is merely a paper prophet. But if he persists in planning on the basis of old-fashioned economics he is not just naive. He is dangerous. A community relying on him must meet disaster. Without effective research on these new economic constraints, political independence in Scotland or elsewhere would achieve very little.

As distinct from research, the immediate economic task of the Scottish government will be to get rid of industrial colonisation. The solution, as Servan-Schreiber pointed out, will not be to put a wall up against the foreign industrialist, far less to eject him; he might even be encouraged to import his research and development activities. The solution is to provide stimuli for our native industry, particularly resource-based industries. Of course, one's industry can be protected by raising tariff walls, but this must be considered a very inferior approach. The dangers are well known. Native industry becomes hidebound and lazy, and so ultimately more easily taken over. The objective must be achieved by encouraging the brains and inventive skill of our own people, by fiscal aid, not merely by reduced taxes. Reduced taxation is distorting, but where 'risk money' is spent on innovations it is not enough simply to allow this as an expense set off against a tax on profit; the tax itself should be cut, in relation to the innovator's whole tax bill and his expenditure on the prototype.

However ingenious the prototype may be, the use of technology is not for the convenience of industrialist or inventor at the expense of the community, but rather for the benefit of both, with the community taking precedence; so any new process must meet rigorous standards for pollution, and be considered socially desirable in the first place.

This innovation, therefore, will frequently need public money and public permission. The community requires assurance that the money is being spent with minimum risk and waste for potential benefit. The resources of Scotland are complex, the opportunities for their development varied; the ramifications of each decision will reach far. No one man, however skilled, can judge, nor should he try. Yet a parliament cannot make decisions unless it has facts before it and a clear statement of the options offered and where each will lead the community. There might be two bodies to provide this information.

The first could be a direct advisory one, a science policy committee, of experts on the technical resources and applications of our Figure 1, who would have to evaluate the results of the different options presented by the problem in question, and clearly lay their findings before parliament. The purpose of this Science Committee, indeed of science itself, is certainly not to allow scientists ultimate command: its purpose is to bring scientists into government – for only scientists know how little is really known – as a check on the unwary enthusiasm of politicians. Then parliament, aware of the options and alerted to the desires of their electors, has its task of judging the whole situation.

The current Scottish Council 'Oceanspan Project' illustrates the point. Even before it was much more than a bright idea, certainly before its implications could be considered, vote-hungry politicians were anxious to put ore terminals at Hunterston and petrochemical complexes along the Ayrshire coast.

Informed protests were violent. There is now too much at stake to allow this sort of 'point-to-point' planning to occur as sudden response to whipped-up desires; but nothing can be done with whatever haphazard control there is, still situated down at Westminster. A deep water anchorage of the calibre of the Firth of Clyde, a facility almost unique in Europe, is a fact; but it may not be logical – for the balanced development of the Scottish community – to make it into an industrial complex.

There is always this choice of futures for a community, and since its betterment lies in the skill with which it deploys its resources, territorial, social and psychological, there must be planning. To assist long-term planning there should be another body, an institute along the lines of Hucsan Ozbekhan's 'Lookout' Institute in California, whose function would be 'to conceive of possible futures; to create standards of comparison between possible futures; to define ways of getting such possible futures by means of those physical, human, intellectual and political resources the current situation permits to be estimated'.[4]

We must in other words keep a variety of technological options open, and keep ourselves ready to assess their impact. Jantsch[5] has shown that 'function-oriented' thinking is coming to the fore. Are we to plan a road system for 1990 without at the same time watching for changes in means of transport or in social attitudes? Of course not, if we are accustomed to the evolutionary approach; yet this was not done at all before 1960. From what we can judge from the defensive preparations of the oil companies, alternatives to internal combustion engines may soon be on their way for urban transport. In other words, product-oriented thinking (as about cars) obscures the view into the future. Function-oriented thinking (as about transportation) keeps the entire gamut of technological decisions open. For the same reason VTOL air travel may well render obsolete the viciously wasteful aerodromes now planned for conventional craft. How often do cumbrously-conceived undertakings roll to expensive conclusions well after they are seen to be out of date? A large unbalanced nation cannot coordinate itself in time to adjust to new trends of thought; England has been failing notably in this for some time. An active new country is particularly suited to take advantage of such insight, and its expert groups will alert it to the possibilities. Finland's remarkable recent captures in advanced European shipbuilding are a case in point.

In small comprehensible communities such as Finland or Scotland it is much easier for the members to control both parliament and bodies such as these: to see they do their job and do not hide behind a welter of sub-committees or a smokescreen of regionalism. Only by such cooperation between expert and layman (and every layman is expert in his own field) can the several technological options open for development of a community be thrashed out; their means of achievement, their immediate repercussions, and a prediction of their outcome. And this must mean on all counts – territorial, social and psychological – not just industrial. To employ jargon one might say we have a chance, by developing such forecasting, of 'optimising the anticipations'. This is the late twentieth-century way of saying life is what we make it; in such a Scottish community we would still have that chance. The patent lack of such a chance today among the vast blundering forces of the 'British' industrial crisis, with their erratic, uncoordinated or dubiously honest 'persuaders' is only too clear; which is why the best opt out or get out.

The Scottish parliament will have much to do: when it dissolved itself in 1707 to rescue its community from irresistible economic aggression by a vigorous self-confident neighbour, it little realised how soon the picture would change. Today that self-confidence has gone from the south, whose current blackmail is a sign of despair. The new Scottish Parliament can put its own house in order, and will negotiate the freest possible interchange between its own community and the others in the world. Part of this interchange will be the provision of expertise to aid the less-developed countries – or those pathologically over-developed; Scotland's contribution to a global equality. There is no reason why we should not be one of the stars by which less-fortunate nations choose to set their new course.

PART III

1. Scotland Today

In Part I we saw how large communities committed to expansionist industrial economics are merely attempting to stave off an inevitable reckoning with the biological requirements for continued existence on this globe; and how in the process they are naturally incurring severe damage to man at his most vulnerable point, that most delicate feeler of evolution so sensitive to any mistake – the communal spirit of modern human society. Among the examples we quoted England. In Part II we saw how small communities could more easily put themselves back on to the right path, providing they could keep off surrounding pressures by a sufficient degree of self-government; we gave Scotland as a particular example among many.

Now we will examine in Section 1 of this Part III, those pressures as they manifest themselves when one of these large disordered communities seeks, for temporary relief, to subdue and so flood into a neighbouring smaller, potentially-salvable, community which has not yet been able to free itself by self-government. There are several examples in the world today, but the most obvious one, subject of much comment abroad, is that concerning Scotland and England. This is the one we shall discuss.

This book, as the reader can judge, is in no sense a manifesto either for or against any existing political party, or against any 'national community' as we can describe that term. Its only brief is to emphasise the dangers of biological imbalance to all communities and to show that certain smaller communities can escape these dangers before, and more easily than, large ones and having once escaped may help to extricate the larger communities from their impasse. Consistent with speeding global removal of imbalance, therefore, we offer in Section 2 some practical suggestions whereby members of a threatened national community can help to frustrate the expansionist aggression of a neighbour which by historical accident has already achieved a tight governmental grip on their fundamental resources. Again, we take the example of Scotland and England.

For the example chosen, we shall not dwell on historical events leading to the Treaty of Union, which, as we noted above, was the only practical means at the time of preserving the Scottish community from the growing effects of its expanding neighbour's economic blackmail. The means could be justified, for the Scottish community, in exchange for its parliament, could theoretically participate equally in England's expansionist policy; and expansion at that time of course incurred no obvious dangers. Besides, a certain degree of self-respect remained; the Treaty contained ambiguous 'escape' clauses by which a good conscience – that precious psychological resource – should be preserved, if not nourished, in Scotland.

Nor will we dwell on the dubious ways in which that Treaty was 'interpreted' or simply ignored by the larger partner in the days when England was confident and successful. Much has been written concerning these past errors and experiences; far too much for the health of a community whose present salvation depends on working for its future. We are interested only in this present condition; when economic blackmail now once again threatens Scotland, but this time confused and vulnerable, from a Treaty co-signatory obsessed and not knowing what it does. Scottish Home Rule politicians may invoke some cancelling clause; Westminster may declare it irrelevant. From our biological point of view it is clear that a Scottish community cannot possibly assist either itself or a struggling England by acquiescing in submergence; then both would drown. Whereas, if Scotland, now buoyant with a new promise, fights off these clutches, she stands every chance of being able to render salutary support to her misguided neighbour in the future. England, to be reborn, needs a Free Scotland. Now, to the analysis (see Figure 2, overleaf).

We wrote of examining the pressures exerted on a community before its engulfment. Figure 2 is one way of tabulating the position. We have again three boxes, and we have labelled them Numerical Aggression, Industrial Colonialism and Cultural Indoctrination. The boxes are obviously inter-related: without Cultural Indoctrination, for example, Industrial Colonialism – bringing in its train Numerical Aggression – would not be permitted to develop; and the results of Numerical Aggression add their local power to Cultural Indoctrination, and provide further excuse for Industrial Colonialism. But because we are not setting out in Figure 2 the logical development of a single balanced community as we did in Figure 1, but rather attempting to unravel here the present attempts to obliterate the balance, the contents of the boxes are not so clearly differentiable. Does televised commercial persuasion from London count as 'industrial colonialism' or 'numerical aggression', or as 'cultural indoctrination'? Or as all three? But the names or classification need not matter; let the arrangement in Figure 2 stand. As long as it brings our attention to these various factors, their cause, the reason and their danger to our future, it will be of value.

The boxes inter-relate to form what we have termed the central government's policy for preventing a Scottish future. They are not, of course, exclusive to Scotland. They operate all round the world. Other communities today recognize their threat well enough, Canada, Hungary, Poland and Czechoslovakia, for example, and to a greater extent, the persistent Basques, Bretons and Welsh; but few communities recognisably *national* by our definition are as tightly gripped as Scotland. Cuba, at some cost, has largely diverted them; Iceland, Norway and Finland overcame them some time ago, and the 'new nations' of the old overseas British Empire did so more recently. But those communities which have not yet escaped are regarded by the 'surrounding pressures' as troublesome im-

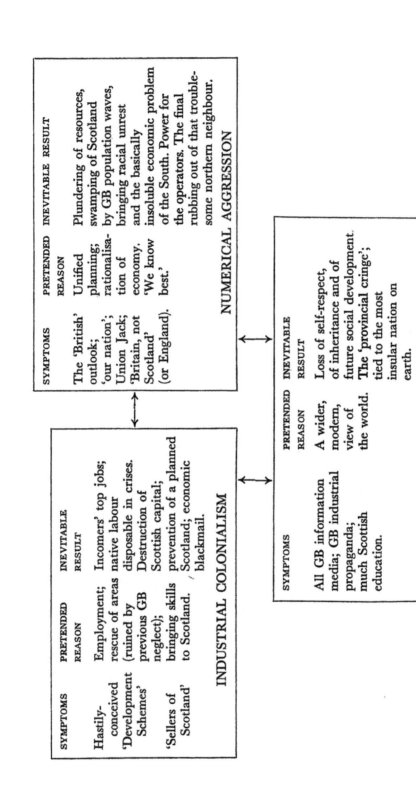

Fig. 2 The inter-relationships of London's policy for preventing a Scottish future.

pediments to their expansion, who by their existence not only get in the way physically, but suggest dangerous alternatives, psychologically, to the politico-economic hierarchies of their aggressors.

In beginning to examine these boxes separately, let us see how this particular 'central government' in London administers the awkwardness of its independently-thinking neighbour; and this will take us to the manifestations enclosed in our first box.

Numerical aggression
This stemmed from the incorporation of the Scottish parliament as a minority in the Westminster one. It followed increasing interference from the South with Scottish financial enterprises at home and abroad; and then in 1703 the Scots Parliament, wishing to avoid entanglement of its people in English wars of expansion overseas, enacted that no Sovereign could declare war on Scotland's behalf, save with the consent of the Scots Parliament. This was rather too much. Obviously, to a healthy imperialistic England an independent peaceable neighbour was intolerable. So all the very considerable skill in diplomacy and espionage of 'perfidious Albion' at that time (Defoe had a long way to go before decaying to Burgess and Philby) was devoted to persuading the Scottish politicians to join the bigger firm.

Every stop was pulled out, including the familiar Highland *versus* Lowland and Catholic *versus* Protestant (both still in today's repertoire). Even so, it was a near thing, the debate lasting three months and including Lord Belhaven's celebrated address, which among other forecasts contained:

> 'I think I see a free and independent Kingdom, delivering up that which all the world hath been fighting for since the days of Nimrod . . . to wit, the power to manage their own affairs by themselves without the assistance and counsel of any other.
>
> 'I think I see . . . a petty English taxman receiving more

homage and respect than was formerly paid to their Mc-Laellein Mhors.

'I think I see the valiant and gallant soldiery petitioning ... while their old regiments are broken.

'I think I see the honest and industrious tradesman loaded with new taxes and impositions ... petitioning for encouragement for his manufactories and answered by counter-petitions.'

How many recent instances come to mind? Lord Belhaven seems relevant now at the end of the day. As Mr Wellington Koo remarked on a similar occasion when a London treaty was outlasting its historical usefulness, the setting sun is darkened with the wings of chickens coming home to roost.

Nevertheless, London's diplomatic skill, at that time, compels admiration. As Professor Iain Henderson records:[6]

'The Scottish Parliament allowed the Union Commissioners to be chosen by the Queen, knowing full well that the monarch's choice would be influenced by her English advisers. Can anyone imagine the English Parliament (today) allowing itself to be represented at the Brussels negotiations (for the Common Market) not by Mr Heath, but by the nominees of General de Gaulle? ... It is almost impossible to explain to Englishmen why Scotsmen did what they did in 1707. For in that year they did what no responsible Englishman, however restrained his nationalism, would ever dream of doing without federal safeguards. They committed the representatives of their country to being a permanent minority in the Parliament of a foreign country.'

Naturally, the supporters of the Treaty dared make no reference back to the Scottish people, who to this day are torn between anger at being taken in and a feeling of duty to those ancient signatures. And every form of psychological assault on our community has subsequently concentrated on that split: producing our three social embarrassments – emotional nationalism, emotional unionism and, creeping moistly in between, the provincial cringe.

As promised, we shall not dwell on the past. But it is sufficiently clear how the present numerical aggression in government derives from that remarkable arrangement of 1707.

For one of the most tedious aspects of being an English member of Parliament is the amount of Scottish business that has to be dealt with. One of the frustrations of representing a Scottish constituency is that Westminster never affords enough time to deal with Scottish matters. The government frequently, as with the recent Divorce Law Reform Bill, enacts the law for England and leaves Scotland to adjust her legislation as best she can through a parallel bill for Scotland brought up by a private member; private members' bills have failed to get second readings on technical grounds, simply because not enough English MPs have been interested enough to attend. Presumably by this means the Scots will see in time the stupidity of attempting to develop their once renowed legal system. But the government takes care never to risk this 'independence' when the issue involves spending anything in Scotland or collecting Scottish revenue and resources, for these are necessary weapons of aggression.

The present curiously patronising government of Scotland has evolved partly from lack of interest and partly to provide enough self-administration to placate requests for devolution. Any changes have been the up-dating of existing systems, rather than creation of new bodies, such as that Scottish Convention hastily proposed by Edward Heath in 1969. Once in power, Heath's government of course equally hastily relegated the Convention farther and farther down the parliamentary timetable; and presumably only another upsurge in SNP votes will cause its reappearance. The instrument of administration is the Scottish Office, headed by the Secretary of State for Scotland.

The power of the Secretary is considerable; the *Daily Record* calculated he had some 3,000 or more appointments in his pocket, at salaries of up to £7,000. Yet, this man, for all his power, is not elected by the people of Scotland. He is an MP on the side of whatever government is 'in', and appointed by the Prime Minister. He is in the Cabinet, but not the 'Inner Cabinet'. He is only consulted 'when Scotland is involved', and by this mechanism one partner to the Union is largely excluded from the major part of the Union's business. On the credit side, it is stated that St Andrew's House, Edinburgh, where most of the administration is performed, is the best bureaucracy in Europe. This bright spot in a depressing picture is of course because St Andrew's House deals with a small national community of a comprehensible size; and in its cage we get a glimpse of what a Scottish Scotland, planned as in Part II, could call upon:

> 'Officials in such sections as housing and local government know the 31 county clerks as individuals, the men in the Department of Education and the Inspectors have a definite picture of the conditions and atmosphere . . . while the Department of Agriculture can describe the practices on individual farms. It is a help in this respect that the majority [still?] of the administrative class of civil servants . . . are Scotsmen educated at Scottish schools and universities.'[7]

But this evocation of an apparently Scandinavian common-sense will not do. We are in a cage, remember. This administrative unit must follow the governmental master. It is a strictly internal body. Though the Department of Agriculture for Scotland can issue a quarantine certificate on a batch of our potatoes for export, it is not allowed to communicate directly abroad; that must be done through 'the Ministry in London'. It cannot decide in favour of Scots farmers, for example, who were prohibited from exporting potatoes in 1970 at a time when prices were good on the continent; the potatoes were ordered down south, through 'the Ministry in London'.

So that, despite its good intentions, St Andrew's House is

effectively hamstrung and its innovation confined to the most trivial matters. Consequently it receives blame for the governmental mistakes it must carry out and, as the Royal Commission on Scottish Affairs, 1952/4 (48) points out, its very obedience obscures for the Scottish elector the small degree of independence he is permitted.

The Secretary of State himself has a most unenviable job. He must satisfy his Prime Minister, or lose office. If he neglects Scottish opinion too much, he will lose his government votes in Scotland; yet if he responds too much to distinctively Scottish opinion he will also upset his chief. His success or failure may be only remotely attributable to his own efforts. The rise in unemployment in Scotland during the last Labour administration was not due directly to Mr William Ross; they arose from policies pursued by his own and previous governments. Yet he and his assistants spent much time explaining to the world how Scotland was being benefited by, and could only be benefited by, this government in London. No wonder it is a servile role that can attract few likeable politicians and even fewer able ones. Not since Tom Johnstone has a Secretary of State earned admiration in Scotland. In office when the war ended, he was responsible for such independent organisations as the North of Scotland Hydro Board, with its proud motto *Neart nam Gleann,* Power from the Glens, and for persuading the government of the day to send the National Engineering Laboratory to East Kilbride. He brought a little self-respect to his task and to his people.

Clearly, whatever his ability the Secretary of State is of considerable value to the government, who can shelter behind him as he behind them. Whereas an English minister is openly responsible for what goes on in his charge and may readily be challenged in Parliament, the Secretary of State can get away with a good deal more because of the less time and interest given to Scottish matters in the House. As he is the Prime Minister's man, his public decisions will never run counter to Government policy; there are no options open to him.

And, of course, the latest move to reorganise local government into nine top-tier authorities will place power more neatly in the government's hands through the Secretary of State, who now has his own representatives near to local level. Outnumbered in Westminster, the Scottish community becomes closer now to being outnumbered in its own burghs and villages.

That is one way numerical aggression affects our government. How does it affect our people?

It is commonly assumed that Scotland is an empty land. Yet it is three times more densely populated than the world average and three and a half times that of the USA, today reckoned to be populated enough. This population has remained virtually static since the last war, even though live births exceed deaths by six per 1,000. A position of astonishing instability exists; huge numbers emigrate, while others immigrate.

Table 1

Population in relation to workers and dependants

Source: *Abstract of Regional Statistics*, HMSO

In 1967	*Scotland %*	*England %*
Under working age	26·0	23·1
Working age	59·5	61·3
Over working age	14·5	15·6
Males 64		
Females 59		

Government terminology is naturally designed to mask the appalling depredation of Scotland's psychological and social resources. Official figures never deal with *actual* emigration, only the *net* difference between emigration and immigration The breakdown was once made by the Scottish Development Department, whose interim report[8] was withheld from publication; it must have been too embarrassing. An unpublished copy was, however, smuggled out and its information has

Table 2

Gross movement between England (and Wales) and Scotland 1956–68

Source: Scottish Development Department

Year	From England and Wales	To England and Wales	Net loss to England and Wales
1956–57	33,300	44,900	− 11,500
1957–58	34,500	44,700	− 10,100
1958–59	33,300	45,600	− 12,300
1959–60	33,300	57,000	− 19,700
1960–61	35,900	62,300	− 26,400
1961–62	37,600	59,100	− 21,500
1962–63	37,500	59,400	− 21,900
1963–64	38,200	62,800	− 24,000
1964–65	39,900	62,600	− 22,800
1965–66	44,200	66,300	− 22,100
1966–67	46,900	63,900	− 16,900
1967–68	48,000	61,400	− 13,400
	462,000	685,200	−223,200
	= 8·95% of Scottish population	= 13·2% of Scottish population	

never been officially denied. In the twelve years (1956–68) immigrants amounted to 9 per cent of the Scottish population, emigrants to 13 per cent; 462,000 persons moved from England into Scotland (see Tables 1–4). In the year 1966/67, 102,900 people left Scotland, some 2·2 per cent of the total population *in one year*. Yet in Czechoslovakia the authorities are desperately anxious at the loss of half that number over the last two years from a much larger population. One disturbing feature of the emigration is that it is 'heavily concentrated in the younger (15 to 34 years) age groups (55 per cent net loss) . . . 64 per cent of the loss occurs among the economically active, who make up 50 per cent of the population'.[9] The work force is denuded in numbers and quality; of those remaining a higher proportion is unemployed than in England. It can be calculated that if the present tide of English immigration simply stays at its present level, the Scottish-born inhabitants will have dropped by 1980 to only 75 per cent; and unless they can regain their psychological initiative they will be dominated by the economically better-equipped and active incomers. Of course, the trend will be immeasurably strengthened by Government plans for overspill.

Table 3   Source: Scottish Development Department

Gross movement home and abroad in the year 1966–67

Immigration from England & Wales	46,900
Immigration from overseas	11,850
	58,750 = 1·13% of Scottish population
Emigration out of Scotland to rest of UK	63,900
Emigration overseas	39,000
	102,900 = 2·18% of Scottish population

The catastrophic erosion of the community's people and spirit is clear enough. And this loss of people has also an economic consequence, which we shall now examine.

Even those countries possessing enormous natural resources, such as Brazil and Australia, know they cannot prosper without modern technology, which in turn depends on an adequate supply of skills. What they lack, they arrange to import by encouraging the immigration of skilled workers. In Scotland the situation is almost reversed. Though it lacks creative people, due to previous earlier losses, the country has adequate resources but a surfeit of unused skills. The skills cannot be used because Scotland is not run (as in Part II) in a technologically modern way for the benefit of her own community, but – by the mechanism we have just described – is tacked on to a failing English economic system. Scotland's potential may be measured in terms of her loss of skills.

Though Scotland has produced her David Humes and Adam Smiths, it is in science and medicine that her contribution – in sheer numbers as well as in quality – has been most remarkable. From 1931–51 the Glasgow Royal Technical College alone produced 10 per cent of all the technologists in Britain.[10] Even in 1967 Scotland supplied 12·7 per cent of all applied science first graduates; of these, 75 per cent had been Scots-domiciled. But it does their country little good. They mostly leave to benefit other nations, and their own only by repute.

The skilled young man forced to leave Scotland because he cannot find himself a worthwhile job at home finds himself elsewhere a valued import. Australia, for example, put a value of £11,000 on a fresh high-school leaver as far back as 1968; this excludes what had been spent on him, and measures only his value to the Australian economy. In Scotland the sum of education, allowances, free meals and so on might be some £3,000 for a child who stayed here long enough to get a Higher Leaving Certificate. So the exported youth is taking with him £14,000 worth of Scotland to another country, because his own is not allowed to gear itself to use him.

The Economist calculated the loss of a first class university graduate was equivalent to a capital loss of £80,000.[11] Since in Scotland at that time a university honours education cost around £4,200, we see the total outlay, school and university was some £7,000. In this way of export business the receiving community gets that £80,000 worth and the donor community loses its £7,000. . . . From emigrant figures published for each skilled group by the Scottish Development Department and using our own estimate of values (certainly low) the authors have calculated that the value of the Scots who have emigrated over 1956–68 reaches £3,800 million.

This continual draining away of the cream has been likened by Robertson[12] to the over-hunting of a valuable animal, such as the Antarctic fur seal. Obviously one cannot so easily evaluate the possible extinction of native talent in the Scottish community. We cannot overlook a long-term genetic effect. Moreover, the increasing drain of these people over the last hundred years markedly reduces the psychological resources of our community and leaves us shockingly open to the effects of more overt numerical aggression. Table 6 gives the first quantitative evaluation of the preferential emigration of the best. The sample refers to University graduates. Qualitatively one has the impression it applies to all levels of education.

Table 4

Value of emigrated workers from Scotland to England 1956/68

Source: Scottish Development Department

Type	% of Emigrants	Number	£ Value/Head (minimum)	£M Total Value
Professional	7	48,000	20,000	960
Non-Manual others	30	205,000	8,000	1,640
Skilled Manual	24	166,000	5,000	820
Semi-skilled	18	123,000	3,000	369
Rest	21	173,000	?	?
				£3,789 M.

J. P. Mackintosh, MP for Berwick, showed[13] that the percentage of Scots entering important UK industrial and government activities was low compared with their numerical strength. He supposed this reflected inferior training and environment, but R. H. S. Robertson has made a much more thorough analysis.[14] Going back to 1600, he has analysed the collective Scottish 'genius'.

It seems that in terms of able and gifted men and women Scotland was at her peak in 1815, a century after the Union. In his analysis, Robertson first looked into those facts that caused such a flowering of genius as occurred during the

Table 5

First employment of Scottish domiciled graduates of Scottish Universities who left Scotland for first employment or further education, expressed as a number and percentage

FOR YEAR 1968/69

DISCIPLINE	MEN					WOMEN			
	Higher Degree		1st Hons.	2nd Hons.	Other	Higher Degree		1st Hons.	2nd Hons.	Other

DISCIPLINE	Higher Degree	1st Hons.	2nd Hons.	Other	Higher Degree	1st Hons.	2nd Hons.	Other
Arts	8 / 17%	17 / 50%	50 / 23%	25 / 13.5%	—	7 / 50%	60 / 19%	50 / 8.6%
Social Science	6 / 16%	10 / 47%	77 / 23%	55 / 12.7%	1 / 20%	3 / 43%	26 / 14%	37 / 8%
Science	61 / 48%	39 / 36%	123 / 30.0%	82 / 18.3%	5 / 25%	4 / 14%	28 / 16%	31 / 11%
Applied Science	30 / 23%	33 / 49%	96 / 35%	75 / 24%	—	—	3	1

Total of 1st Class and Higher Degree = 224

'golden age' from 1795 to 1850, a time 'when Edinburgh (and Scotland) was a force in European civilisation'.[15]

'In this general advance Scotland was vitally involved. And Scots talent found particularly marked outlet in directing the new theoretical gains to practical technological ends ... Scotland had become more prosperous than ever before, and there was a great literary and artistic revival.'[16]

Quantifying the subsequent decline is, of course, more difficult. If Scots are leaving, who are taking their place? The influx from without the Scottish community will, we have seen, even at the present rate, result in only 75 per cent of the 1980 population having been born in Scotland. This is an astonishing proportion. In the reaction of the Scots, irritation has yielded to helplessness punctuated by outbursts of anger.

An analysis of the distribution of educated people in Scotland[17] showed that it contained 87,000 fewer such people in the dynamic 20-34 age group than the UK average suggested.

Aggression on the industrial and cultural fronts is geared to aid this apparent surrender; even the use of 'England' by the BBC has been changed to 'Britain', to convince us we are an illogically-reacting region. Surely our industry needs the incomers, for we are crying out for jobs, aren't we? (The blatant propagation of such a *non-sequitur* ensures that – by the Goebbels principle of 'the bigger the lie . . .' – it is equally blatantly accepted.) And surely we are not . . . 'racialist'? This dreadful word, as shocking to the contemporary English puritan as 'sex' was to the Victorian, hides an equal shame in both of them; and like any other taboo-word, it commands great potential political power, as both Mr Powell and the unnameables who operate behind decent front-men like Mr Bonham-Carter, very well know. No, unlike the English community whose present 'racial' self-abasement is in equally unpleasing proportion to their previous 'racial' arrogance and proves their long-standing and hysterical obsession with the subject – unlike them, Scotland whether at home or in its representatives abroad within or without the British empire, has never been accused of, and has therefore no reason to fear, the guilt of 'racialism'.

Rather do we stand by the right of a community, in Scotland as in, say, Algeria, Scandinavia or Central Africa, to develop along balanced lines with free use of its inherited resources. And because these resources, territorial, industrial and – in this issue of numerical aggression by influx – psychological, are being senselessly and irreparably destroyed to do England no good and to prevent us having a future, we object. No man, because his father (or his grandmother) was a MacSporran, is any better a *man* than an incomer to Scotland; he may often be worse. But as a member of the Scottish community he is potentially – even in his present cultural poverty – more valuable in rebuilding his one community in a bewildered Europe than an uprooted and usually rootless incomer ever could be. If the Scottish community were in better balance, or if the influx were less, then nothing but good could come of some considerable degree of population exchange. But at this present crisis, and on such a scale, and disguised by such nastiness, it is sheer numerical aggression, and to be resisted as such.

Scotland does not need immigrants at any but a specialised level, yet her borders are wide open to all, whether they be beneficial or disastrous to her development. No independent country in the world does, or can afford to, permit such imperilling of the future of everyone within – immigrants along with the rest. We dare not experiment here with overspill. The disastrous results of one such venture still smoulders in Ulster; if England wishes – as she now seems bound – to repeat the gamble, in England, we wish her the appropriate luck; but *here* our modern planned Scotland will arise from our own resources.

So that is what we term 'numerical aggression', from Secretary of State down to the 'British' overspill he is administering. We recognise it by its insistence, necessarily, on a 'British' outlook; by the unrolling of plans on a 'British' basis, a 'unification' in a divided world. We shall be treating of its economic effect under the box 'Industrial Colonialism'. Let us here look briefly at a classic example, the Tayside Report of 1970.[18]

This report was born, among others, from the realisation in Whitehall that the English population was growing too fast and that contingency plans must be laid for overspill. Tayside, a region of exceptional beauty and resource, was one selected reception area; Tayside, which would have been one of the key places for careful development by a Scottish government to restore the balance of the Scottish community, was to be engulfed by the all-too-familiar out-of-date economic expansion of an uncontrollable 'British' community. Through the Secretary of State a group of economists and planners were given the financially rewarding job of putting forward designs, and reasons, for an initial mere doubling of the population. Wary of reactions, the central government demanded no more at first – only later, as we shall see, did one of its ministers, confused by shame, let the cat out of the bag. To soothe community feelings, this committee gathered to itself various local dignitaries, who were given the soft sell. The Tayside Plan was to them as a scheme – the only scheme – that would bring work to their towns and villages. Of course, it was nothing of the sort. It was a scheme whereby the present inhabitants of Tayside would pay for the development of the infrastructure to meet the needs of someone else's increase in population; ratepayers from Arbroath to Blairgowrie would pay for these people's sewers, roadways and site developments. The social problems created by mass immigration of the rootless were not apparently considered: or perhaps they were, and were judged a worthwhile exchange for the eradication of those psychological resources still so inconveniently strong – as election figures for home rule candidates showed – in the independently-thinking north-east of Scotland.

There was little demonstration, and certainly no likelihood, that local people would benefit from this influx. Their petty amenities such as golf and fishing would cost more, their motoring jaunts be reduced to the fuming queues of England. More bitingly, the local traders and business men would not reap the 'increased markets': for these would – as in all previous cases – be exploited only by the big firms of the south, interlinked with the property developers. The expansionist economy of England, however destructive to its own community and that of Scotland, naturally does some people a power of financial good. The property and insurance groups that had been cheaply buying over the hinterland of Tayside (and they are busy all over Scotland, for they know what is planned) would certainly see the fruit of their schemes.

We must emphasise that it is *not* the development of Tayside to take new people and industries that is condemned; what is condemned is its development for a basically *unplannable* English overspill of factories and people at the expense of the Scottish community's *plannable* future. The destruction of local capital resources may be considered later. Here we stress the destruction of our basic biological resources, land and people.

So sharp was reaction, the community instinctively recognising that the future of its children was being sold away, that the central government hastily used 'Scottish sentiment' as a defence: those who opposed the plan would be 'simply against the development of Scotland', traitors who wished to see Scots remaining unemployed (why they *were* unemployed in the first place was naturally not dwelt upon). However, since 'Scottish sentiment' may induce inconvenient community feeling and so strengthen the victim's psychological resources, the proper castration had to be performed; Scottish sentiment had to be directed to its own destruction. The local MP and Government Minister, George Thomson, was picked out for this humiliating task. He did it in London in June 1970, and very badly, to his credit. In him did those three embarrassing weaknesses collide violently – emotional nationalism, emotional unionism, and the provincial cringe? Out of the mêlée one gathered that Scotland's national gesture was to assimilate England's excess population....

This unfortunately premature revelation of the real reason for plans such as that of Tayside produced a stunned resentment; and since then the preparations have continued less obviously.

To aid, therefore, such numerical aggression, much must be made of a 'balanced' British economy; what this 'balance' may be, compared to the basic biological one we have discussed in Parts I and II, can naturally never be explained. Also, to convince the victims that their own resistance (rather than the aggression itself) is selfish and reactionary (another good Goebbels device), any 'nationalism' must be played down or, as the insufficiently-agile Mr Thomson tried to do, be turned inside out. There are sneers at 'Scottish' nationalism and mild deprecations of the perceptibly-rising English nationalism (for the Englishman is no fool and does not usually carry a pose far enough – especially on his own soil – to endanger himself). Numerical aggression is, however, *itself* nationalism. And as there is good and bad nationalism, and as we believe from contemporary examples that a small advanced community is capable only of good nationalism and large unbalanced ones – USSR, China, USA – usually of bad aggressive nationalism, therefore we must class this trend as a peculiarly undesirable form of 'British' nationalism; a nationalism false as well as overbearing, for there is demonstrably no single viable 'British' community and this nationalism, so far from arising beneath from the people, is advertised heavily from above for the economic reasons we have already discussed.

To this sad end of plundering the resources of the most viable national community in the island and persuading it out of existence, of eradicating the one-time Union partner, comes the once-proud Union flag; it hangs now grimily over each new regional centre or imported factory, like a pin on a salesman's chart. Those who were courageous enough to risk death beside their national flag in past wars – and those politicians who were practical enough to provide that flag – might well consider how the respective communities of Scotland and England, freed by a new Treaty from the mutual destruction of numerical aggression, could work side by side under the saltire of St Andrew and the clean bright flag of (alas, and perhaps symbolically, he is no longer a saint, according to the Vatican) George.

Of all the resources of a community – or a world – the psychological resource is the greatest. Numerical aggression destroys it in victim and in aggressor.

Industrial colonialism
Now for another aspect of the 'pressures exerted', that contained in the box 'Industrial Colonialism'. This is intimately linked with numerical aggression, but here let us consider how the threatened community can be further embarrassed by the imposition of factories or businesses controlled from without. At first sight these 'gifts' appear wholly beneficial. Yet even the most under-developed communities, like the new African nations, seeking such aid because of their under-development, are wary of its effects, Zambia for example being particularly wide-awake; whilst welcoming their share of the mutual benefits of foreign capital investment, they hedge these foreign companies or foreign state enterprises about with careful restrictions. They know that the USA, China, USSR, West Germany, any 'aider', is naturally interested in perpetuating the need for his aid.

Scotland is not under-developed but developed one-sidedly, to an imbalance. It is most unlikely that the colonising firms are interested in correcting that imbalance, and it is only too evident that the central government is desirous of extending it so irreversibly that the Scottish community will be pushed past any chance of a future. So no wonder that industrial colonisation under the central government is almost wholly pernicious.

How is this perniciousness disguised, or impudently passed off as positively beneficial to our community?

Firstly, Scotland must be shown to be a depressed area; this is not difficult. Secondly, this depression must be shown to be *inherent*, not of course a result of 'British' imbalance; it must be shown that Scotland cannot pay for herself; as a corollary, the fact that Scotland has resources for a high standard of living and certainly would pay for herself under self-government must be firmly kept hidden and if necessary denied. Thirdly, this 'inherent' depression must be proclaimed curable only by the injection of capital, skills and labour from the south.

How does the aggressor show that its victim is inherently unable to pay for itself? By disguising the victim's natural resources and by rendering inaccessible figures for the actual cash flow in and out. By claiming that the victim therefore requires subsidies. And by quoting the (undemonstrated) existence of these past subsidies under bad conditions as proof that the victim will never in the future under any circumstances be able to pay for itself. Remarkably, several people appear to accept such a performance as firm proof; would they cancel a picnic tomorrow if it had been raining yesterday? Presumably so.

One does not, when establishing a company, expect to show an instant profit. Indeed, shareholders expect to lay out their money for several years before a notable return begins. One establishes a company on the basis of considered prospects. A director is the more willing to invest his money because he knows he is in control. If, as one political party avers, Scotland already pays more to the Exchequer than she receives, then the prospectus for the new Scotland is very rosy indeed. But it would still be perfectly valid to launch this enterprise even if Scotland were not today viable, as long as it could be demonstrated that her long-term prospects were better with self-government than remaining with England. As Parts I and II indicate that prospects on self-government *are* very much better, then presumably it is now incumbent on the unionists to prove the continuing value of the 1707 agreement.

Note that there is no dispute about Scotland's ability to *survive* independently; the dispute concerns how *well* she can survive. The Conservative Party's Scottish Constitutional Committee admitted that 'Any nation can exist independently at a price'; but they took care to add 'We are convinced that the price which Scotland would have to pay, economically, politically and socially, would be unacceptable to the Scottish people'. The basis of this 'conviction' appears emotional. Throughout 326 paragraphs this committee[19] never once adduced evidence which might refute the arguments of this book or which bore on any constructive *policy* for a new Scotland. It was either incapable of imagining a modern approach or was too satisfied with its temporary personal niches in the present fiscal confusion.

For confusion it is. No figures exist to clear it up. Experts themselves admit this. The Glasgow Chamber of Commerce, wishing to ascertain the effects of independence on trade, offered *The Economist*'s Intelligence Unit £3,000 to come up with the answer. A month later this Unit announced that basic information was simply not available to make the assessment. . . . Yet in spite of this lack of information the Conservative committee could confidently speak of 'conviction'. The United Kingdom itself is well served with statistical data but Scotland separately is not, and we have no means of making a scientific evaluation of Scotland's present economic position relative to England.

One attempt has been to analyse the income-expenditure relationship between the two partners. This has given widely different answers because key values such as income tax can only be guessed at. One of the most useful of these partial analyses is by Gavin McCrone, now with the Secretary of State. He concludes that for 1967/8 Scotland received more than she paid in. Other studies on different bases arrive at different conclusions. No one has compared the small change in such balance sheets with the massive capital tied up in the present control and utilisation of Scottish human and physical resources by others. No one has made an assessment of Scotland's potential given control of her own resources: surely the whole point of self-government!

McCrone's book *Scotland's Future: The Economics of Nationalism*[20] is gravely mistitled; it does not deal with the future of the present Scotland, still less with that awaiting a readjusted independent community. He writes (with our interpolations in brackets):

> 'The problem of the Scottish economy ... is its structure [Agree] ... industries whose employment must on any reasonable economic assumptions decline still account for too large a proportion of the country's economic activity [Agree] ... these are not problems of an under-developed agricultural region, nor is there any parallel between Scotland and the Scandinavian countries [We agree; as we have pointed out at length, no parallel at present can exist, for the free Scandinavian countries are *already* balanced; but an independent Scotland, unlike England, can join them in that balance].... Scotland is suffering from a singularly British economic sickness. [Only too true! This is our whole point; but, McCrone does not take the logical step of separation from 'Britain' and its sickness: he falls back on the time-dishonoured stalling strategy of British regionalism, although using tactics in themselves often admirable] ... if this problem is to be overcome it requires a continuing effort to re-adapt the Scottish economy around new forms of economic activity [Agree]. The labour force must be retrained [Agree] or if necessary the population must be regrouped around centres which are suitable for the development of modern industry. [This sounds ominously familiar]. Inducements of all kinds are necessary to attract new firms to Scotland [i.e. Industrial Colonialism] and to encourage the expansion of those that are already there.'

McCrone has made no attempt to cleave the problem to its roots – the basic resources, the biological balance, the driving spirit available to the community; there is no conception that Scotland should be administered *for the good of its community by its community*. Only its administration for the British economy by the British economy. Always this desolating acceptance of a basic lack of responsibility, of delegation to others. If our book has any purpose it is to point out that the survival of Scotland – indeed of modern society – cannot lie in this sterilised anti-human way of thinking. Compare McCrone's thesis with that of R. H. S. Robertson.[14] Both are consistent with most of the existing facts. McCrone simply sees a need to restructure the present economy, and is under the impression that this can be achieved from without Scotland. Robertson explains also why the situation arose and suggests a more lasting means of success, a restructuring from within; his is the dynamic evolutionary approach.

McCrone, whose heart, if not his reading, has been in the right place, tells us we would start to get better if only we could start to get well: '. . . it would be an entirely healthy development if the Scots came to realise that the origin of many of their problems and solutions rests with themselves.' A sadly indisputable truism. But how can they, when they are belaboured by such confusion? Lord Provost Fitzgerald of Dundee has apparently given up trying; at a lunch in London in June 1970 to 'sell' his part of Scotland he said 'The future of Tayside lies in the one square mile of the City of London'. Mr Hugh Stenhouse, Conservative Party Treasurer and Glasgow business man, stated that he was 'not a Scottish Nationalist, but a Scottish realist', presumably firmly basing his realism on the inadequate information we saw available to his party's committee.

With such 'realism' in the 'leaders' of our business, it is no wonder that the central government feels free to complete the economic servitude of the Scottish community. To give the government its due, it is rather embarrassed by the eager surrender of so many Scots business men: the President of the Board of Trade remarked wistfully in 1960 'I should like to see more promotion by Scots of the virtue of investing in Scotland'. But its most recent Secretary of State has no time for such sportsmanship; Campbell goes at the sitting bird with both barrels – 'The Highlands and Islands offer the developing industrialist a real alternative to the congested areas of Southern Britain.'[21]

That is the drive behind industrial colonialism, and again it takes a certain type of Scotsman to reveal it through the unpleasant role of a 'collaborator'; no doubt, in honest belief, as the honest but unfortunate and reprehensible Quisling was used for cruder methods of destroying his community's pride. There is, of course, no real governmental desire for Scottish business men to think for themselves. All they need do is to queue, like our Lord Provost of Dundee, at the London canteen and the rations will be handed out. They may even, with the right connections, be promoted to the kitchen staff, like Sir Alec Douglas-Home who, as with the inimitable smirk of an upgraded institutional cook, remarked of the less successful aspirants, 'Scotsmen know which side their bread is buttered'.[22] Imagine the impossibility of such a remark, delivered in the USA, being made there by an Englishman about his own community: certain 'Scots', we shall call them victims of cultural indoctrination, try to bury their own shame in the overall humiliation of Scotland. That honest Englishman George Borrow called them 'foxes that have lost their tails'. Sir Alec will, however, agree that we are only joking.

So, despite the lack of figures for the present economic standing of Scotland, we gather we are handed our butter from the south: we are being subsidised. Many Scottish MPs claim Scottish resources are inherently so meagre we will always have to be subsidised this way. We trust they cannot all be foxes without tails, so this pessimism must be based on genuine ignorance, imperfect education regarding their country; but it will not improve the morale of the community that elected them, and they should be responsible enough to remedy it. However, suppose we were being subsidised just now; most colonies, however well-endowed, suffer this way. Nation 'A' overcomes nation 'B', commands and exploits its resources, employing its inhabitants to do so. Since nation 'B' is in a state of patronage with low income, little of its money is liberated at home. It is clearly dependent on nation 'A' for all improvement. But how true is it to say 'B' is 'subsidised' by 'A'?

We do not wish to dwell on this confused picture, but an example of a central government's tactics may be instructive. A few days before the first Scottish by-election after the home rule success at Hamilton, the government issued a 'budget' for Scotland,[23] naturally to be regarded as a propaganda exercise. Scotland was shown to be 'subsidised'. Had it been otherwise, the budget would not have been issued and some other tactic more favourable to government objectives would have been selected. The Treasury 'estimated' that for 1967/8, on current account, Scotland was £130M in the red, and had borrowed three times as much capital per head as had England. Other budgets, such as those of the *Financial Times*, McCrone's, the *Scotsman*, and that of the SNP, all showed less imbalance or none at all. There are three ways of establishing this sort of budget: (a) a study of payment flow between Scotland and the London Exchequer; (b) allocation of respective benefits enjoyed by each, from existing revenue and spending patterns; and (c) assessment of the financial position of a modern independent Scotland. For (a) statistics are very poor; (b) requires intelligent guesses. As (c) dare not be considered, the Treasury compiled an unsatisfactory hybrid of (a) and (b). Their 'budget' did not show, as income generated in Scotland, any tax paid either by employees of nationalised industries or by those myriad branches in Scotland owned by outside firms. It attributed absurd unprofitability to Scottish-based companies in its assessment of Corporation Tax. In dealing with revenues from oil and motor taxes and from excise duties, it used figures which it later had to concede were false. But although these subsequent corrections would put Scotland safely back in the black, they came too late. That whole government exercise can be regarded as malevolent, but it was successful. Subsequent recantation in small print by the government will not erase the impression of millions that they each get £48 a year more out of the government than do the downtrodden English. . . . It is possibly worth looking at some of the figures.

Expenditure per head on university education is given as £8·05 in Scotland, but a mere £5·4 in England. Of course, the cost of the 23 per cent of English students at Scottish universities should have been deducted, bringing the Scottish figure down to £6·20 and the English up to £5·6 (there are very few Scottish students at English universities). There is no mention of the vast flow of Scottish graduates to England. The first-class honours graduates and higher-degree people alone can be valued at £18M for that year, and even putting the rest at half their value one sees a loss of another £23M. This would leave an expenditure deficit of £600,000 and a capital transfer to England of £41M.

Funds paid out on social security benefits are higher at £52·6 per head as against the UK figure of £50·9; but obviously if Scotland had not a higher degree of unemployment than the rest of the UK, this figure would be less than the average. Housing is shown to require £38 a head in Scotland as against £19 in England; but on allowing for tax rebate on mortgages, this disparity largely vanishes, for the proportion of owner-occupied houses is so much higher in England. Agricultural support is almost twice as much in Scotland, but this is due to the food price policy of the government. Scotland takes most of the forestry funds, but of course Scotland lends the Commission most of their land except – predictably – for research and administration.

Now let us leave this essentially sterile discussion; it has bogged down too many expositions. It deals with the present, unknown situation in which he who shouts loudest is heard;

and the central government, having the most powerful voice, is concerned to keep it that way. No budget has ever been postulated for condition (c) above – a modernised independent Scotland. But the wind of change is blowing; more people are turning to such a refreshingly obvious approach. This book is a first product of the stirrings. It will not be the last.

Having ensured enough confusion to be able to assume the necessity for subsidies, the government can then tell the Scottish community that its only real hope is subsidy extended to the form of industrial settlement. Hence the incentives to development in Scotland. These various schemes, besides being good propaganda to demonstrate London's concern for her ailing Region, have the effect not of assisting Scottish economy but of rendering it less and less able to stand on its own feet.

Let us take a recent example. In February 1971 Scottish unemployment rose to 118,000, highest since the Depression. Suddenly the government announced a £100M plan to help out. It was hailed with eulogistic gratitude – '£100M bonanza' and so on. The idea was to give a 30 per cent subsidy to the payroll of new ventures over the first three years. The alleged reason was to alleviate unemployment by the help of the more competent south (where Rolls-Royce became bankrupt the next day), apparently a most generous gesture. But was it? The amount laid out would save on unemployment benefit for the government. But for Scotland it would generate labour-intensive industries in a country whose vital problem was to prevent the export of her gifted and skilled people and use them to create her industry from within. But, worse of all and hardly credible in its effrontery, this subsidy was not to be available to Scottish firms, *but only to incoming enterprises*. It was a direct subsidy on economic aggression, an invitation to industrial colonisation. English capital was given access to cheap Scottish labour. Scottish capital and initiative were penalised. And as well as attacking Scottish industry, it assaults our psychological resources. The long-term effects will be serious as a new generation of wage-slaves emerges in industrial Scotland, their inevitable disillusionment providing wonderful opportunities for exploiters of social unrest; all ensured by a right-wing government.

Not helpful psychologically, either, is the practice of special houses for incoming key workers in these industries. Built at the expense of the Scottish tax-payer, himself stretched to provide for his own badly-housed and ill-paid compatriots, these houses are usually better than, and set well away from, the 'native quarters'. Such blunders will be as harmful in the end to the central government as they are at present to the Scottish community.

Another unbalancing effect of incentives, this time in a rural area, is seen in the overlarge pulp mill at Fort William. To work economically it must ingest incessantly a vast amount of pole-size timber; and the West Highland landscape's visual resources and its ecological balance are being ruined for generations by the frantic destruction of seral oak and birch woods, hill pasture and good arable land (there are many instances), so that mile after square mile of sitka spruce monocultures can be planted; and once planted, be sprayed and fertilised by air until after thirty years they can be clear-felled by mobile gangs. Everything must go – balanced development of local industry, fishing, farming and resilient afforestation – to fill the maw of this Moloch, huge enough for the endless forests of British Columbia and Alaska but ludicrous for the precarious timber economy of Scotland; and all on the assumption that this paper combine will continue to require wood-products made up there at that rate, and that the sorely-harassed soils of the West Highlands will withstand such cropping. It is the thoughtless all-destroying Sheep Boom again. The Highlands provoke such economic excesses as a reaction to the sentimental indulgences of their defenders; industrial colonialism, for which despite Mr Campbell's sales talk they are not suited, is regarded by some as peculiarly 'realistic'. In the context of a 'regional' Scotland, the Highlands would be far better left as a Recreation area for Britain; in an independent Scotland their partial industrialisation is certainly desirable, but requires social sensitivity, biological knowledge and a sense of economic proportion, all of which were notoriously absent from the Corpach adventure or the Invergordon smelter.

Bombarded with assurances of their ill-health, staggering under forceful injections of 'help', the Scottish working community, top and bottom, is losing heart. The *Scotsman*'s industrial correspondent observed in 1961,[24] 'The great question, still unanswered, is whether the Scots have found the industrial recipe necessary to produce economic dynamism'. The answer is there for anyone to read. Economic dynamism has no existence apart from dynamic people. If the government perpetuates a system that encourages the export of dynamic people from Scotland, one can hardly be surprised if the result is to leave behind an unambitious or stagnant society. By claiming the answer is unknown, those in power can indulge in any kind of irresponsibility. For example, as protest against unemployment Scottish Trade Union leaders planned a one day strike; one can hardly visualise anything more calculated to prevent reflation in Scotland and benefit the South. Five months before, the Secretary of State had joyfully proclaimed that 'The steady improvement which is taking place (in the Scottish economy) is something we should be shouting about'. Yet some hours later, questioned, he admitted a loss of 153,000 jobs under his administration; a year later losses reached record levels – even the self-employed were being driven out.[25]

Another blow is to be administered by the Conservative government of the day. Returning to the nostalgia of free enterprise, Mr John Davies, Minister of Technology, stated that they would no longer feather-bed non-viable industries. The approach was to be a return to raw capitalism. With so much of her land in the hands of limited liability companies and so much of her industry closed down or taken over, Scotland, the pessimists say, is ready for complete assimilation. Naked capitalism is not an opportunity for Scots to exercise their self-confidence, their courage or even their ability at home. It is simply a chance for those who own anything to get a better price by selling out and getting out.

It is painful to record such dreariness, but it is an inevitable part of industrial colonialism all over the world; the degradation of the colonised – before their eventual uprising. Look in the newspapers. Not a week passes but some body in Scotland is out and about 'selling' Scotland. These people are not to be confused with trade missions, which are Scots business men out looking for orders and deserve our best wishes. We refer to those cap-in-hand delegations of the Scottish Council and various Trade Unions and Town Councils touting round the world offering the labour and land of the Scottish community as if it were so much compost. The branch factory mentality.

Yes, the sole result of government incentives has indeed been to convert Scotland into an industrial colony. By 1968 one in ten industrial workers was working for a US firm or subsidiary, and one in twenty for an English firm. By 1970 there were only eight companies employing over 10,000 people whose headquarters were in Scotland. One of these, Burmah Oil, has its main activities outside Scotland. Only two, Distillers and Coates, Patons Ltd, had significant international connections. The 120 top Scottish companies in 1969[26] collectively employed 330,000 people (35 per cent), exported £153M (28 per cent) and made £83·4M profit on a total capital of £1,167M, an average return of 7·14 per cent. Though one or two traditional companies made a profit per employee that IBM would be proud of, by and large the figures showed that although under its present crippling disadvantages in a 'British' economy Scottish native industry does surprisingly well, it is all too vulnerable – with this organised aggression – to take-over, assimilation or extinction.

The further effects of industrial colonialism imperil the Scottish skilled workers as well as their top executives. Notoriously, branch factories are staffed at the higher grades by people from the parent factory; notoriously, these factories are the first to be closed in recession or to pay men off. We all know local examples. Conditions in England already send more and more skilled English people to employment in Scotland (an ironical reverse of our brain drain) and with the inrush of overspill industry competition for top jobs puts the advantage very much on the English side; for it is natural to appoint your own men from the parent factories, and in response to such firms' advertisements, ten times as many English as Scots are available for each job.

The Scottish Council (Development and Industry) admitted in its Centralisation report,

> 'As the positions of authority within industry diminish in number, as opportunities for employment in Scotland at the top of the tree diminish – whether for people in general management or in finance or in marketing or in research – so also does the vigour and drive of the whole public life of the community run down.'

Earlier it remarks of those who do the take-overs,

> 'It is quite unrealistic in human terms to expect the decision-makers in a highly centralised organisation to attach great importance to the regional social implications of their operation.'

Remember well those words. They do not spell the doom of the Scottish community. They spell the doom of the present economic system. Read Part I again. Man is not outside biology; the community will certainly prevail. But it will prevail sooner when it is informed. Meanwhile, we suffer things like the crassly insensitive behaviour of the British Sugar Corporation directors over the intended closure of their Cupar factory; and Imperial Tobacco, those 'absentee landlords' of Smedley's food factory in Blairgowrie, retracted their threats of closure only after strong local protests led by the Provost, a man rather more realistic than the down-climbing 'decision-makers' and, in spite of being a Scottish Nationalist, certainly more realistic than our quoted Mr Hugh Stenhouse.

Even the thriving Scottish electronics industry, being largely in colonial branch factories of some twenty foreign companies, was cut back in a minor recession for this reason; only one firm, which had brought research and development activities with it into Scotland, continued to expand – on good Carnegie principles – when the rest were contracting. As well as ourselves in previous Parts of this book, Dr David Simpson, Scottish National Party economist, has stressed the necessity for concentrating on these 'R & D' facilities in Scotland.

Let us end our examples for this box on industrial colonialism with the brighter side. In spite of all these constraints, Edinburgh still remains a very considerable financial centre. There are still two Scottish-based, Scottish-owned and Scottish-operated independent banks controlling over £1,000M of assets. There are eight life assurance offices with 1968 share values of almost £500M. Standard Life is the largest mutual company in the UK. Two new merchant banks have recently opened up. Though controlled from London, there are considerable Unit Trust Funds handled from Charlotte Square, of which the Save and Prosper group capitalises on Scottish sentiment with their Scots-shares, Scotbits and so on. The Scottish Stock Exchange in 1969 had a turnover of £430M and floated £216M of new issue. One substantial insurance company, General Accident, survives. On a *per capita* basis, Edinburgh runs London a close second as a financial centre. The estimated book assets of these various institutions amounted in 1969 to some £6,945M.

Another activity that still flourishes is agriculture. It does not necessarily do so in a business sense, but it contributes enormously to the country's welfare. The output of Scottish agriculture was £220M in 1968, which with a work force of 54,000 makes an output per man of some £4,000 : comparing very favourably with the rest of the world and even with many industrial units. Agriculture is the largest single industry still largely in Scottish hands, even though many farmers are tenants of land owned from without. In spite of the continued decline in the viability of farming, the price of farm land has risen all over Scotland as people and institutions seek to own land as a hedge against inflation.

Let us now summarise. However shifting and obscured the details may be, the overall picture is clear. Again we see that (a) the Scottish community is suffering from the aggression of the central government; (b) this aggression, whether intentional or not, is inevitably incurred by the economic union of the two communities; (c) the aggression will persist as regional imbalance even if complete destruction of the Scottish communal consciousness occurs (as in the example of English Tyneside); (d) removal of aggression requires economic command by a Scottish government; and (e) from Parts I and II it is safely predictable that the separated Scottish economy will soon catch up and then overhaul the far less manœuvrable English one.

We must remind ourselves of the purpose of an economic union, such as that shared by England and Scotland today. As McCrone says, 'It is to concentrate resources, whether they be of capital or labour, in such a way that they yield the best return.' It raises the immediate questions: concentrate whose resources, and where? and yield the best return for whom? and what about those great determinants, psychological and social resources? McCrone himself adds,

> 'Should it ever become clear that the Union is working consistently to Scotland's disadvantage, and that politically and economically she would be better off as a separate state, then many Scots might come to feel that their elected representatives should repeal the 1707 Act.'

We have tried to demonstrate that the Union is now in fact working consistently to Scotland's disadvantage; that economically and politically she would soon be better off as a separate state than she is now; that to remain attached to the economic system of 'Britain' is to be engulfed in the plight of England without being able to help her. This plight is the continuous threat of economic blackmail – export or starve – and an inevitable financial and social disaster. Industrial colonialism is not of course sheer malevolence on England's part; it is a natural, desperate, symptom of her own doomed economy. We must get clear of this drowning clutch and help from the shore. The Director-General of the National Economic Development Office in London admitted in October 1970 that if Britain did not achieve high growth within the next two years 'then we will drop out of the rank of major industrial powers . . . once we cannot compete . . . the whole basis of our way of life collapses and we can no longer sustain over fifty million people on our present resources'.

Scotland has her own options. We owe it to our children to take them.

Cultural indoctrination
The third box of Figure 2 illustrates the most serious of the 'surrounding pressures' a threatened community has to withstand. Cultural indoctrination attacks its very spirit. Numerical aggression may dilute that spirit, industrial colonialism mortgage it; but when these assaults fall back on themselves – as they eventually must – a community that still holds on to its spirit can shake itself clear and go forward into the future, leaving the aggressor wrestling with his own internal troubles. Past colonies have done this; present satellites are waiting to do it.

But equally the converse holds. If this spirit, which is the self-respect engendered and sustained by the psychological resources examined in Part I, is weakened, then any aggression penetrates deeper. It may even destroy the community, and so involve its members in the eventual debacle of its aggressor.

Therefore cultural indoctrination is the most dangerous of all pressures. Moreover, it cannot be so easily recognised by its victims as are the floods of incomers rushing for jobs and houses, or the colonies of usurping factories. Like the wooden horse it may even be welcomed into Troy as an amusing diversion.

In wartime, psychological assault is stepped up to more obvious levels; we laugh at the excesses of the enemy's propaganda. We smile, subsequently, at those of our own. We do not undervalue either, at the time. After a war, the process spills over into commercial exploitation. Many have written on it – Vance Packard in *The Hidden Persuaders* and Hoggart, for example; both have demonstrated such aggression against resistance to sales and sales-values.

Governments, too, employ this weapon in peace time, in the form we have termed cultural indoctrination; it is used especially by those industrially committed nations to weaken resistance against their obsessive expansion. Our news media pour scorn on the methods used by say, Moscow against Czecho-Slovakia, Mao against India; as we move more Left through these media, we encounter criticism of the US psychological pressures, of the pushing of the American way of life against Canada and the Latin neighbours. The more sophisticated news media examine playfully from time to time various psychological assaults on contrived 'British' resistance groups, the Consumers, or Conservationists, even – if they are right wing and have secure circulation – on 'Teenagers' and 'Students'. How rarely do they examine, or admit the existence of, any assaults on the genuinely living communities within the British Isles?

Why? Because a genuine community, if informed of them, could repel these assaults. And such assaults are in grim earnest. On their success depends numerical aggression and industrial colonialism : necessary for their advertisers' expanding economy and so for the present existence of the news media; necessary for the central government's power. There seem to be few newspapers or radio programmes that dare to challenge nowadays the authority of such a government. Do not confuse 'government' with the different liveries its members wear from time to time – Conservative or Labour. You will not find today, as in the two previous centuries, influential news sheets critically examining other forms of government. The British form of government, certainly until recently quite good, is apparently considered not only the best, but one that cannot now be improved upon: sure evidence of political stagnation and community decay. Stagnation means decay, but the new world is, as ever, only for those who can evolve. The 'British' balancing feat, one concludes, is becoming too precarious for that sort of enquiry to be permitted. So we must tread carefully here.

At present most resistance to cultural indoctrination from a community such as Scotland is quite accidental, because the community has no conscious awareness either of its own psychological resources or the forces working to destroy them. Later, in the second section of Part III, we shall discuss effective means of neutralising those forces. Here we shall try to summarise them. Again we must emphasise that we choose the London government's assault on Scotland merely as one example of a global trend, wherein certain large nations committed to a cul-de-sac in economic evolution enter the final phase and begin to struggle more wildly. Much of our analysis would apply to those communities neither acceptedly national nor independent, such as the gallant Basques, attacked, fortunately, by the two rivals, France and Spain; or to those communities, strongly national but temporarily constrained, such as Hungary, attacked by Moscow; to those national and independent, such as Iceland, attacked from the US base at Keflavik (where in some parts it is not permitted to speak Icelandic); to Viet-Nam, attacked by China (the US attack is so clumsy as to be a source of spiritual strength); or to others such as Australia or the Scandinavians, keeping a clear and wary eye. It is a most interesting phase of history, this preliminary cultural sparring for selection as members of the World State. We must steer Scotland through to acceptance not just because we can give that far-distant state some good advice and example, but because if we don't steer through, the next few Scottish generations will become engulfed in one of the most unpleasant political, social and racial upsets of Western Europe; when England – to quote again our official London source – 'can no longer sustain over' (well over, by then) 'fifty million people on our present resources'. It is bad enough being next to it; we must never let our children become part of it.

By drawing attention to their plight we do not wish to embarrass members of the central government; we understand why they wish it hidden from their people (who never had it so good). By deploring their policy of cultural indoctrination we do not assume them evil men; we understand the exigencies of increasing pressure, the pain of a swelling abcess. Certainly, by attacking their so understandable plans we do not wish to attack *them*. And if they consider a stronger community spirit in Scotland means a weaker 'Britain', it does not prove our thesis blameworthy – it merely proves it true.

Cultural indoctrination is easiest with a common language between aggressor and victim; easier still if the victim's language and with it so much of his verbal heritage, is replaced by the aggressors. The Teutonic Scandinavian countries have, after separation, been careful to develop the distinguishing characteristics of their related languages. Again, Finland's astonishing achievement in technology, and in spiritual triumph over both Russia and the US, is greatly due to the transformation of native Finnish from a socially despised gibberish to an authoritative organ of Government; scarcely any young people now habitually speak the once-proud, but psychologically-provincialising, Swedish. No wonder the desperate young Welsh paint out the English road signs that march ever farther into all they have of a Finnish future. Englishmen themselves have been known to complain bitterly of 'creeping Americanism'.

Scotland has English as the official language; that in itself is no worry. French-speaking Quebec fears New York, not

Paris. But it does make the aggressions easier. We would not expect all Scots to speak Gaelic instead. Our community has evolved to speak English, or Scots-English, and survived; and like Australia, or the early United States, it has a promising enough future with its practical variety of English. But let us beware of the assumptions that can be so glibly slipped in; that it always spoke English, and thereby allowing that there is no difference. And the 'we are a single community, aren't we?' follow through of psychological assault, which ironically enough used to be made on the US by England.

The first official language of the recognisable Scottish Kingdom was Gaelic. Then that shrewd English politician Queen Margaret, that *sair sanct*, brought along her language and by continual persuasion, right down to the English Ambassador Sir Ralph Sadleir at the Reformation, ensured its adoption throughout the government. Never, as usual, subjection by arms. Invading bands of Northumbrians had always been driven back, language and all; so quickly that one of their kings, Edwin, had hardly time to cool his backside on Edinburgh's rock. But hundreds of years later he did achieve a peaceful subjection, for the facts were 'persuaded' and Duneideann, the ridge fort (Dineidyn in Brythonic), was provincialised into Edwin's burgh; and the so-called Scottish capital held to be derived from the northern seat of an English kinglet. Despite the labour of scholars like Watson or Jackson and others on the 'Edin' names of Gaeldom, Edwin reigns on in the history books of many Highland schools. Indoctrination starts early and less clumsily today; a hundred years ago Gaelic had often to be thrashed out of these pupils.

Cultural indoctrination uses all available media; radio, television, commercial advertising, governmental announcements and many others besides school books. Should you not believe this, go to Russia, or better, one of its satellites, with enough of the language to understand; you will be chilled by the rain of propaganda there, because the very unfamiliarity of its doctrine makes it obvious to you. The inhabitants notice it less, even accept it (yet they are intelligent people). Go to Canada similarly and be chilled or amused (depending on your conditioning) by the US persuasion there, from candy 'packs' to gift libraries. The Canadians do not notice it so much (yet they are intelligent people). In Scotland there appears little of this undermining. But perform a similar experiment. Go and live abroad for some time in a civilised country free of aggression. Or, easier and just as effective, stay at home; but cut off wireless, television, newspapers (except say, *Le Monde*), and read in the resources of Scottish literature and history, in Scottish ecology and geography and more widely in any world culture but the English. (You may of course go to your job provided it is not part of the indoctrination complex!) Then after some weeks, switch on the mass media, read them, listen to them; and you will be astounded by the raw obvious blast of ceaseless propaganda, of half-truths, of glib assurances, all, whether deliberately aimed or passively accepted, tending to the one end: the obliteration of the Scottish sense of community, the denial of its resources. But nobody else notices (yet they are intelligent people). Within a few days of exposure, your conditioning will re-assert itself, and your indignation fade; so powerful are these forces, and so insidious. You will require another spell of analytical thinking.

For analysis we can divide the attack of Cultural Indoctrination into two prongs: Divide and Rule. The first prong encourages factionalism, the second, provincialism.

Factionalism
Here the victim must be split, so that he wars within himself. The community must be convinced it is inherently unstable psychologically – an exact parallel to the economic attack – and unfit to rule itself. The device is much used by imperial governments. The classic and tragic local example is the 'British' legacy to Ireland; but we will touch that pitch no more here. Scotland has several potential lines of split, and all have been attacked; but, thanks to the practical nature of

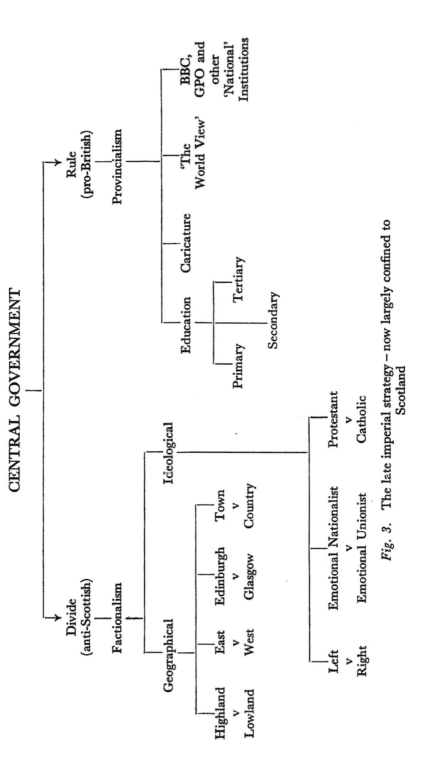

Fig. 3. The late imperial strategy – now largely confined to Scotland

our community, none has yielded much. They may be classified as follows:

(a) Geographical. Here the most promising approach used to be Highlands *versus* Lowlands, but as habitual Gaelic has now vanished from the entire mainland, a unity of language once again strengthens the Scottish community from such an attack. The ignorance of Scottish history engendered by provincialism allows use to be made, for example, of the help given by a Lord of the Isles to the English against the Scottish king; but all histories are full of such internationally-minded princelings. This attack, like many of the factional ones, is illogical; if Highland is so different from Lowland that Scotland could not thrive as a unit, then how can 'Britain', compounded of demonstrably more divergent and self-aware communities, be claimed permanently viable on the same grounds?

East *versus* West is an alternative geographical approach. In the north, the theme 'West-coaster' *versus* the allegedly more provident 'East-coaster' would split relatively few people. The more populated south allows the promising theme of Glasgow *versus* Edinburgh. Here the necessary friendly competition – nicely balanced and salted in the immediate past between Glasgow's greater industrial fame and Edinburgh's historical and administrative status – has been blighted by a recent denigration of Edinburgh. She has been docked of influence in the regional authority shufflings while Glasgow expands to become a Southern Highland capital; Edinburgh has no prospect of a motorway link with the London central administration which has been Glasgow's for some time; east-coast rail links are threatened, to relegate her to a branch line from the London-Glasgow route. Putting Edinburgh in her place as a north-easterly British provincial town could be an effective check against Scotland's – and so Glasgow's – future prosperity. Good psychological attacking this – to open a split and simultaneously damage Scotland's pride; for however Anglified and East windy Edinburgh may be, she is a dangerously suitable rock round which Scottish self-government could crystallise.

Other more general geographical devices, such as Town *versus* Countryside, are involved when required by all central governments or interested parties during expansion drives. If we only remember their existence, examples of their application against the community spirit of Scotland will be the more readily recognised and exposed.

(b) Ideological. The splits are much more more promising. In politics they have been worked on successfully for many years, though a stiffening resistance is now noticeable. But superficially, the Labour *versus* Tory farce still, like the rivalry of football teams, holds a public carefully awakened to no deeper issues. One may hear the bitter remark, 'they're both the same, there's nothing to choose between them'; and the surly pendulum swings, in confirmation, first one and then the other monotonously into power; only occasionally certain offensive Bills or characters vary the performance. But Labour *versus* Tory, however discredited, still serves its purpose. It still keeps the community divided, however superficially. A Scottish Labour supporter bristles, as he is meant to, at the (somewhat directionless) Tory arrow stuck on his neighbour's car window; a Tory clutches his *Daily Telegraph* (as he is meant to) at the sight of fellow citizens on strike. A right-wing central government tells the Scottish right-wingers that independence would deliver them to a left-wing government; Scottish left-wingers are equally chilled by the solemn converse prediction. Such infantile bogey-bogey still frightens many, despite the fact that every few years the central government, right- or left-wing, anyway hands its Scottish supporters over to a successor of the opposite political hue.

Political thought tends towards what we term the 'Left' or the 'Right'; as Marx and Lenin demonstrated, this polarisation has been intensified by the stress of industrialisation. And, of course, the whole spectrum will be available to an independent Scottish electorate. But, in spite of their platform 'opposition', one can scarcely impute any political thinking along either 'right' or 'left' lines to the present London parties. Labour appears increasingly run by insensitive administration careerists that in Scotland are paid for by a London Office; the Tories by an equally unprepossessing claque of bumptious salesmen; the Liberals, thin as they are, cannot force themselves into the narrow gap between. The electorate, long loyal to the old colours, is understandably restive. But what other choice is there?

In England another duet (on a rather more active plane of deceit), Fascist and Maoist-Trotskyist-anarcho-communist, waits its turn in the wings, fairly well pleased the way the stage is being set; there are no other successors to today's lumbering pair.

In Scotland, of course, there *is* a third effective party, one seeking to steer clear of the mess and promise independence. To stymie this party, the central government must keep wedged open that breach begun in 1707: it must encore our highly dramatic and highly ineffectual duo – the emotional Nationalist and his equally ridiculous Unionist counterpart. In the past, between them they kept the issue of independence safely in Melodrama and out of serious London theatre.

We are familiar with the emotional nationalist. He still lingers on from his heyday in the thirties, political version of the stage Scotchman; a cloak and heather figure who, like the Balmoral Games, at least kept one 'idea' of Scottishness going. He has long outgrown any usefulness, but persists embarrassingly in letters to editors (which because of their entertainment value are invariably printed) and among the ousted fringe of practical separatist politics. For him, love of Scotland necessitates hate of England and Englishmen. The Scotland he loves is that of a tartan sweetie-tin, of battles long ago and always romantically lost. He is, of course, basically defeatist. Hydro-dams and factories rouse him to frenzy not because they might be in the wrong place or in excess, but because they are bad in themselves. He cannot evolve. His equally static partner, the emotional Unionist, is only anxious to differ from him as dramatically as possible; he naturally welcomes every exploitation of his country's resources, and denigrates its past. The emotional Unionist cannot understand why Englishmen, whom he serves so ostentatiously, seem to prefer his more honest partner-in-stupidity.

The audience for this tiresome and mutually-dependent pair is trickling away to more critical discussions of Scotland's need for self-government. Even as puppets, however, we can expect them to be vigorously manipulated for some time yet; any distraction, even such a Punch and Judy show, is of use to those who pick pockets.

Factionalising continues in other, ideological fields. Religious faction-mongering we will not dwell on; we hope the matches struck so provocatively so many years ago across St George's Channel have now blazed a sufficiently conflagatory warning against a central government's meddling in such matters. But as they never appear to learn, we should remain on our guard.

The Landowner *versus* Plain Man theme is dead, now that it is realised that most of the few existing old Scottish 'landowners' are as anxious as we are to save our community; their successors, the invading southern business syndicates and state corporations who are buying us all up are certainly no friends to the Plain Man; but they of course are never revealed as the 'landowners'. Indeed the task of tracing the real owners of Scottish land calls for considerable research.

Provincialism
A community convinced that it is crippled by innate divisions is the more readily provincialised. Assimilation to a powerful, rational and united aggressor seems the only sensible course. Whereas in the past England may have appeared in such a protective role, its present condition scarcely justifies the pose. Committed by abdication of moral leadership to a minor world role, committed by irrational expansion to a disastrous economic future, and committed by greed and

foolishness to the increasing likelihood of social discord on a Chicagoan scale, England no longer offers a consolatory escape. When Big Brother goes off on a Rake's Progress the only salvation is to keep clear until he is repentant enough to profit by your own sane example. But provincialising has eaten deep into Scottish self-respect; only now, when our own resources are being revealed as so obviously superior, do we have our first chance since 1707 to check it.

Provincialising starts in the schools. We do not wish to labour the defects of the schoolbooks, of history which is local English history, the history of the most insular nation on earth; of geography, which mentions, in passing, the infertile northern region of Great Britain; of music which sets 'Here we go gathering nuts in May' against the finest folksong in Europe, perhaps of the world; of poetry which gives us A. A. Milne. There are exceptions, whose fervid reception, alas, only proves the rule. Perhaps part of the trouble lies in the Training Colleges; we all know the teachers of 'elocution' there, whose mouthings of the spurious diphthongs of Middlesex do so much to unsettle the self respect of a good Scots tongue that it begins to whimper and mew like certain recent and notorious Lord Provosts. The bewildered adolescent, assured he is of no community, and that he comes from a land of no resources, with a shameful history and a barbaric culture, is hardly equipped himself for a balanced development; he lingers on in gaucherie or escapes to violence, London or the provincial cringe.

This form of cultural indoctrination – considering the victims' history and culture to be worthless or, better, not considering it at all – was practised by most European imperial powers in their African or Asian colonies as long as they held them. Over the last century the more influential members of the Scottish community, familiar with this technique, have recognised the dangers of a provincialising education on their scions, and sent them south; Oliver Brown has noted that Scotland lost her aristocracy at Flodden, Eton and Harrow.

Provincialising continues in the universities, where some

faculties may be overwhelmingly, staff and and students alike, of English origin and outlook, who regard 'Scottishness' as an amusing foible and any rare department of Scottish studies or literature as a generous concession to local antiquarian interest. Such ignorance, of course, is not likely to be corrected by their Scottish colleagues, most of whom have been culturally emasculated. One speaker in a staff common room referred to the Church of Scotland as a 'Nonconformist sect' and was genuinely puzzled when informed that, if he wished to apply such a description, then in Scotland it rightly belonged to the Church of England. At a neighbouring university a student complained that a Scottish chaplain was an anomaly; the increasing number of English entrants required his replacement by a Church of England priest. We quote these particular instances, not to dwell on theological niceties nor to suggest that a moderate exchange of scholars is anything but beneficial, but to illustrate the astonishing lack in the universities of any sense of the Scottish community: either past, present or future. Obviously we cannot look to them to fulfil their primary role of invigorating the culture of the community in which they lie. Culturally, they are dead; scientifically, they continue to provide evidence of the resources by which the community will be able to revive them.

Scottish spirit is obviously not going to be strengthened much in school or university – except by the inspiration of individual teachers themselves; and probably this is the most effective way, for it canalises both the community feeling of the young and their antipathy to the faceless 'systems'.

Having educated someone away from his community, the next step is to get him to laugh at it – destructively. Make him feel big – he needs to, poor chap – by letting him guffaw at those not so educated. Set them up as caricatures. So the wry procession of stage Scotchmen. Had the kilt and pipes not been commandeered over the last two centuries for brave service in imperial wars, they could scarcely have survived such an alliance with the bawbee and crooked stick; the 'auld Scots bonnet' certainly did not and when, as a recent gimmick, airport workers at Glasgow were asked to wear it, they indignantly refused. Remember it is not the laughter, good humour or decency of many of these willing caricatures we deplore, but their always implicit message that in a 'realistic' modern world the Scottish community is not worth taking seriously. Fortunately, a new generation of entertainers, largely with deadpan Glasgow humour, is putting across exactly the opposite point of view. They are not quite so widely publicised.

But to serious matters. We must take the world view. In so dazzling a prospect, Scotland must be shown to be ludicrously provincial. Again, this method of indoctrination is less easy now. The world is full of eager new communities, respectably 're-nationed', resurrected from the debris of outmoded empires: 80 more since 1945. So that this approach by the central government varies remarkably and incompatibly between bullying by the strong – Domination (you don't exist because you're part of us) and the new bullying by the weak – Appeal (you wouldn't leave us, at this time?); 'Us' always being 'Britain'. Since the rise in separatist votes great care has been taken to play down the use of 'England' north of the Border.

One of the main arguments of this book has been to show that, whereas Scotland and England can be considered as national communities, albeit with different outlooks for the future, 'Britain' is not a community but is either, like 'Europe' in its present connotation, (a) a collection of communities or (b) a trade-name for a ruinously aggressive industrial exploitation by a few great combines.

This book stresses that the undermining of the spirit of national communities (whether in the British Isles or in Europe) by an ultimately-futile industrial aggression (whether disguised as 'Britain' or as a 'Common Market') is the greatest danger modern society is facing. Already its symptoms are clear enough in moral and environmental deterioration. The urgent task of economists is *not* to speed this juggernaut on its course but, with the help of the new technology, to slow it down; and to re-think world economics biologically, so that it becomes a service to the whole life of each community – whereas at the moment the community gives up its life for 'economics'. It is in this context that the current dangers to us in Scotland of a 'British' outlook must be considered. And it is such a context which is made impossible by 'Provincialisation'. So this process must be recognised and attacked in all its forms.

One of its forms is flag-waving and packaging. Exploitation of the Union flag has reached ludicrous depths since the Back Britain pseudo-campaign; we have that heraldic monstrosity daubed on shopping bags, on the greasy rumps of 'British Bacon', on the salacious pubic regions of blowsy women whose opened legs invite us to Visit England (fortunately not Britain, this time). We recorded within the same week, saltires fluttering bravely over Arbroath and the superb white and red flag of St George over a church in a Gloucestershire town. One had the feeling of being back among living practical nations again – among the blue and white Finnish flags along the shipyards in Rauma or the Norwegian colours in the docks at Narvik. There are certain things a living nation does *not* do to its flag; the contemporary prostitution of the Union flag should remind both partners it is time to stand again by their own communities. 'Britain' like the British Empire was a useful historical necessity; to cling to either when the usefulness has fallen to hindrance is unrealistic, and, as the posters demonstrate, degrading.

There are many other provincialising approaches, some of which – like that responsible for discrediting the Scottish Law – we have mentioned in previous sections; others – such as the suppression of the name 'Scotland' in postal addresses (the zip code numbers act as a good excuse for this), come readily to mind. Concerning this last, it is symptomatic that, for example, most scientific journals published in England, change authors' addresses to (say) Glasgow, XYZ 123, United Kingdom or St Andrews, Fife, Great Britain, whereas similar papers accepted by United States journals retain, or even have added

'Scotland' as the country of origin. A book may not legally have inscribed on the title page, 'Published in Scotland'. It must be 'Published in Great Britain' (or 'England'). One may find dozens of such instances increasing in number since the recent vote for separatist candidates, but we will not indulge the reader with them here; it is an illuminating pastime to collect them for oneself.

PART III

2. Resist and Achieve

Now we must briefly consider the reader's role in counteracting and overcoming these pressures; what each of us has to do to help our community to regain and develop its self-respect, so that it can reach a greater degree of independence and provide the rightful future for us and our children. As before, we shall limit ourselves to Scotland, but, as before, we emphasise that these techniques apply to all national communities threatened by the present temporary, but extremely powerful, economic aberration in human society. Our action is vital *now* as the world settles toward the constituent states of its first-ever approximation of unity; it must achieve the most psychologically-viable demarcation of the national communities. If you wish *your* community to be one of these satisfying states, able to progress farther, then you must act for it *now*; its fate is in *your* hands.

Strengthening the community spirit of Scotland is something we can all do, whatever our job, our age, or the rightness or leftness of our political views, or even, to a large extent, the degree of our desire for home rule. Our action is something positive in a society deliberately trivialised, something to increase the variety and interest of our lives, to give meaning once again to ourselves and to our fellows, and meaning to the land we have been told is only ours to sell.

Perhaps the womenfolk will grasp it first. Biologically more perceptive than their men, they know that Scotland is our family home in the world, they know what home means to their children and how pitiful is the aggressive insecurity of homeless children; they know how to defend their house from the smooth operators, who are only too willing to come in and look after it for them, or who know a chap who will give them a good price for it. The canny housewife recognises the slick salesman, the plausible landlord, whatever his pitch. Family homes are hard to come by in the world today; they are beyond price. The boys want to get you out, to keep you moving, at their beck and call. Keep an eye on them and on your menfolk – don't let them sell you out, or mortgage the place for a sure-thing streetcorner bet.

Of course, the task is a big one, but certainly not too big. Scotland is more than capable of salvaging itself. It is in a better position economically, geographically and even psychologically than many other communities that have won through from collapse. Finland, for example, with long exposed frontiers, deeply penetrated and controlled by Sweden and Russia, its language driven to the woods as unprintable gibberish, its people at one time reduced to about a quarter of a million; Norway, which ended a 400-year night of economic exploitation and cultural blackmail by a Sweden and Denmark enjoying *their* imperialist day. Both Finland and Norway had 'Unionists' even more unimaginative or downright selfish than our own, yet they burst into fire again from their ashes – and how Europe has benefited from them! The spur to these communities was the same as that now stimulating Scotland – the sudden awareness of what was going to happen to them, everybody, if they did not do something quickly. We have in Scotland enough people with intelligence and perseverance; we have an aim which is biologically sound and therefore practicable. Opposed to these assets is an administration confused, irrational and unstable, both economically and socially. All the basic advantages are on our side. However small our individual action, we are working *with* life, not against it. The cumulative effect will be irresistible. The trash and trivialities will melt away. There is a job to be done. Let us, with all reverence, thank God for the chance to do it.

In essence, action should be: resist the negative, achieve the positive. Resist the erosive measures – listed in Figure 2; preserve and increase the resources – listed in Figure 1. Go over these Figures, until you know them by heart. Add more instances to the lists. Classify every incident you meet or read about by these Figures. Remember you are not being 'political'. Fostering the community spirit of Scotland in the present unstable state of western society is only sheer simple prudence. Nothing can be lost and all may be saved. Remain as right-wing or as left-wing as you are. The job is to achieve a community where Right and Left can mean something positive again – or where both can be clearly shown to be out of date. In the present 'Britain' they are a tedious commercial programme we cannot switch off. And their producers will not read letters. You are not being 'political'; you are not working for a political party, you are working for your society.

Continue resistance with achievement wherever possible. For example, those of us who are experts – and, as we have said, most people are 'experts' in at least one field – should gather together the existing facts of Scottish resources in our own field, or discover new ones. The field can be of any type; supplies of minerals, management of crops, folksong, fishing, literature, design of inshore transport, electronic design. Then you should educate others by spreading information on these resources and explaining their value to the Scottish community and to the world. Show how these resources exist as human resources in a Scottish community and possess their greatest value there, both for its members and for the world; without a Scottish community, one more life-spring has died and humanity's psychological bank is impoverished a little further. At the same time expose as false all denigration or denial of these resources. Expose as imperilling your community any dispersal or destruction of the resources. This kind of work is not idle political theorising; it is practical help for your own society, desperately needed now and easily performed now.

In your work, neither produce nor pass anything shoddy; remember you, as an expert, hold in trust your community's resources; workmanship, self-criticism and communal pride are essential to maintain and improve them, whether you design transistorised circuits, plan housing estates or give the Immortal Memory. And remember that all criticism of present unsatisfactory Scottish affairs must contain a constructive follow up; otherwise you merely, like the whining 'emotional nationalists', sap morale further.

Not only in work or on special occasions can you resist and achieve. Everywhere, at all times, is this possible. For example, never use the terms 'Britain' or 'UK', use 'Scotland' – on letters, posters, books. If you refer to a geographical area, then 'the British Isles'. After the gobbledygook postcodes on your address write 'Scotland', if your mail is crossing the seas or the Border. Only one of our universities – symptomatically the most modern – has grown up; its letterhead is proud and simple – the University of Stirling, Stirling, Scotland. The remainder huddle provincially, secluded in Fife, or hidden behind the stutterings of surely the most expensive and uncouth zip code ever constructed.

If you are in trade or business, then add 'Scotland' after your location; 'Made in Scotland', 'Printed in Scotland' on your labels. And when you are shopping, buy those articles labelled like this, for you know the people who made them are conscious of their community and the need to keep up their standards within it. This is only enlightened self-interest on everyone's part; and if you do find an inferior article labelled 'Made in Scotland', write to the maker at once and tell him that it is inferior. Where you will lead, the rest will follow, so that acknowledgement of their community by name or by flag will mark those concerns which are determined to improve their service and which consider *you* their customer.

Small actions, but cumulatively of enormous significance; the means used by Norway, Finland, by the now-accepted Commonwealth nations to achieve, or regain, their national self-respect. And as well as being constructive, such small things are pleasantly satisfying, discovering instances is rewarding.

Of course, when you say 'this country' make sure that it *is*

Scotland you are referring to; make sure that when the collectively national 'we' is used – as, 'we ought not to allow them to destroy our coastal fisheries' – we are Scots; and not 'we' but 'they' are indulging in East-of-Suez nostalgia, gun-running, European gambling or racialist adventures. Make sure – but tactfully – that other people, too, pick out these distinctions in spite of the wool flung over all our eyes.

Learn about, and be quite ready to admire, the characteristics of the English community, your nearest neighbour – which is now so pathetically anxious to gather up its own broken threads. Avoid the politicians' blanket of generalisation; pick out the clearly individual. Pick out George's Cross from the Union flag, and pick out the Saltire. The Lion Rampant is too tinged with royalty, and royalty, despite the charm and intelligence often displayed by some of its members, has always been shamelessly used by Westminster (or Whitehall) as grease to the political wedge. The hypocrisy of the investiture at Caernarvon Castle served its purpose well; it split the growing spirit of the Welsh community at just the right moment. Anyway, only the Queen is allowed to fly the Lion. The Saltire is Scotland's flag, though unlike the Union flag, it carries tax, and therefore is hardly ever available. (A large consignment ordered from the Crown Colony of Hong Kong was mysteriously 'unable to be manufactured' at the last moment.) When it is available it usually appears in the dismal institutional blue of the Union flag; it should be bright azure, the blue of a spring sky, and the white a brilliant foam white. The Saltire, or St Andrew's Cross, if you incline towards the theological Right, is like all flags a most potent symbol to a community, and you should display it at all suitable times and on all suitable places; as on car windows, bicycles, hotels, petrol stations (good for Burmah Oil! You know where to re-fuel and do not forget to say *why*), campsites, rucksacks, tents. But you cannot get it? Ask for it. In every sports, motor or toy shop. Keep on asking for it. Every Tuesday, say, go to a different shop and ask for a Saltire. In despair the shopkeepers will pester the wholesalers, the wholesalers will tip off the manufacturers; sales will talk. If one out of every two people who voted for the Scottish National Party bought a Saltire, 150,000 would be sold. Buy in some or make them yourself. The Saltire is very easy indeed to make, once you have cloth or paper or the right blue; housewives can run up dozens. And don't be 'provincial'. If you can display several flags at your hotel or in your shop or on your anorak, do so; those of the other communities you have visited, or who have visited you, or who are likely, with increasing tourism, to visit you in the near future – Sweden, Norway, Denmark, France, the United States of America, the community of (St) George. Put your Saltire among their flags – it makes the point better in company. Such flag flying is colourful and enjoyable, it springs from life.

As the Commonwealth Games had to be held in Edinburgh, London needed to think up a poster motif which would be a safe one, without any National nonsense (Scotland being a 'nation', of course, only in the sporting convention of 'gentlemen' and 'players'). They chose the caricature approach: a little man in a bonnet with a crooked stick, called, 'Wee Mannie'. This was going too far. Edinburgh at such times (as over the notorious Royal visit in 1952) tends to remember its eroded self-respect, and protested loudly. These protests achieved ... a real Saltire! They did not achieve the National Anthem. 'Scots Wha Hae' is too honest, rather too relevant these days ('See approach proud Edward's power'). We were fobbed off with the Victorian pop 'Scotland the Brave'; as if England had to play 'White Cliffs of Dover' at its Coronation.

So our course is quite obvious. Fly the flag they fear; call for the anthem they fear. As a general rule, whatever London fears, do. Whatever London offers, refuse. But accept and discard critically, for London often makes mistakes; any little concession, any useless floor-scraping conventions, accept – it is a toe in the door; but watch your toe all the same. Don't look a gift horse in the mouth, but remember it may kick from behind. Let the outline of London's attack, as described earlier, be your guide. Classify each approach and administer the appropriate remedy. The process begins with the morning paper and ends with the last glib television 'news' – or entertainment item. It is instructive to discuss at work the tactics employed by the central government and its collaborators; it is remarkable how ready people are to catch on, how glad they are to clear out the clutter of 'pseudo-events' with which our minds are fed; how wonderful they find it to make the first steps in reconstructing their own community.

The mass media must be utilised. They are used everywhere to erode individual communities in the interests of easy commercialisation. The BBC in Scotland has recently become a talking point for its apparent servility. Why, they even decided to clamp down on that last splinter of nationality – Rugby Union – by cutting out the result of the match at which Scotland defeated England – even with the Prime Minister and Prince Charles watching the game! The England-Wales match was shown in Scotland, instead of the Scotland-France one; and another Scottish international was cut out to show us Cambridge students paddling along by themselves. The most insensitive of Unionists (whether Rugby or Scottish) were righteously indignant – even incredulous – to have their own noses rubbed so cavalierly in the dirt; it made many people realise the current 'British' trend, and watch out for the rather more subtle means employed by the BBC in fields other than sport. Use such instances unsparingly in discussion. Write constructive letters to the BBC Scottish Controller at Glasgow, detailing your criticism (time and programme) and indicating the improvements required. Mere abuse, though refreshing to the giver and possibly deserved by the recipient, is conveniently produced by the BBC as evidence that the writer is a crank. Five hundred sensible letters – praising the good, criticising the bad – would strengthen the hands of the Scots in the BBC; 1,000 would produce an effect at a top desk and 5,000 might result in an improvement. The price of a stamp and ten minutes' thought; keep a stamped addressed enevelope handy and after every particularly nauseating exhibition of provincialism (or after the rarer hearteningly mature approach), write down your views and post them off. There is even an organisation 'Action for Scottish Broadcasting' which can use your help. Local commercial radio will be more sensitive to local criticism, so one can wade in there with great effect; write producer and advertiser.

Newspapers are thirled to their advertisers, or, rarely, are expensive pets of their proprietors. An editor knows what's good for his butter. But even so, a deluge of letters (even if only a selected 'silly' few are published) is not to be ignored. Buy Scottish newspapers (not just those printed in Scotland) and tell the editors why. Most are provincial, one is really dreadful, but its competitor the *Scotsman* tries hard. A good game is to pick out the 'provincial cringe' in this dreadful one – it has so long wallowed underfoot that every snippet of news, however unlikely, is tainted; after a drought which threatened Ayrshire farms with a grassless future, what did we get? 'It was raining heavily yesterday, spoiling the Whitsun holiday in several parts of the country. The South Coast was particularly affected. . . .' They hit it every time, unfailingly. A quite disgusting production, but valuable for discussion groups as an example of the 'perpetual cringe'. The Thomson press is worth praise if only because of the mentality that keeps churning out the Broons; stuck fast, unable to progress, neither does it abandon its wholesome kailyard broth of seventy years ago; significantly, its Head Office flies the Lion Rampant. The *Scotsman*, tight on its master's lead, tries hard; try with it.

Use, in conversation or argument, the points made in this book. Expand or improve them. Encourage thinking about them. Informal local groups will gather to discuss them. Read widely in Scottish literature and history; especially in the latter, allow for the viewpoint of the author, his date and income group. Take every chance of bringing local occasions up to date and beyond; on a town's anniversaries stress the continuity of its role in the Scottish community, and that community's role in a developing world. On Burns's night, encourage modern Scottish poetry, song and music. You will find this approach astonishingly successful: something like relief bursts through with the applause on such occasions – the shell is cracking, you feel the strength of a community gathering with the wonderful realisation that Scotland still exists and will develop. Robust evangelism or quiet demonstration – either method will do to open the eyes of your fellows. They will do the rest. Be quite frank with the innocent hordes brought in by industrial colonialism. Accept them gladly if they wish to work for Scotland, respect their decision if they refuse; in a Scotland of growing community spirit, they will either fall in and feel at home or fall out and drift elsewhere – the simple exploiters certainly will!

The exercise can be applied at any time, examples taken from any source, however unlikely. Here, for instance, are just two items from an extremely unlikely source, a March issue of a horticultural trade journal, *The Gardener's Chronicle*. The first notes that the Swedish timber houses erected by the Forestry Commission at its forestry village, Dalavich, by Loch Awe, and once boosted as an instance of the new Scottish rural communities the Commission was creating, are now standing empty and being turned into tourists' chalets. Why? Because 'economics' has now discovered that, despite its earlier confident predictions, fewer men need after all be used in the forests. (Men, of course, are the least important from this point of view.) Where have the disillusioned members of this short-lived village been dispersed to? To factories, no doubt, to build more machinery to displace other of the 'new rural workers'. A good example here of the negative to resist; the break up of an apparently genuine attempt to re-establish a living indigenous community in the Highlands, and the substitution of that most precarious and parasitic tourist economy – the bane of the Highlands since its first appearance as deer-stalking and the bane of any region granted no other, more stable, pursuits. Of course, any attempt to resist, by letter or action, would be met by factionalism – town *versus* country probably; you are selfishly denying the use of those houses to people wanting a holiday ... so be prepared for that 'answer'.

In the same issue of this obscure periodical is a note on Finnpeat, peat supplied by a Finnish company, prepacked and available in several mixtures, specifically pre-fertilised and treated for different types of plants, including one for treeseedlings. Each purchaser is assured of a free consultant service for his particular needs. With the growing shortage of loam (it is all being built over or destroyed by 'chemical' agriculture) horticulturalists are turning to soil-less composts, usually peatsand-fertiliser compositions. The vast Scottish peat-fields remain sporadically exploited and peat is certainly in no sense scientifically extracted, treated or marketed here. Yet Finland, for reasons we have already explained, is driving ahead and exporting with great success. Here is a 'positive' achieved. Think of the peat beds behind abandoned Dalavich....

So, even from this most unpromising source, a casually-picked-up trade journal in a barber's shop, we find sufficient examples to set a discussion group alight all night, and ideas for action growing. Just imagine what a newspaper, or one of those pressurised television programmes will furnish.

One not unrelated incident from the other side of Scotland: a group of skilled but disillusioned Forestry Commission workers were made to see the light. They were tackled at a local crisis. Their new District Officer had arrived full of the 'new economics'. Sitka spruce was *the* pulp tree. Everything must be Sitka. There were some fifty acres of hybrid larch (a previous Commission craze), well tended by these men from planting ten years ago and now beginning to surge proudly. He had these trees poisoned and underplanted with Sitka. Fake 'economics' apart, what appalling insensitivity to the morale of his men! But men are secondary. One of them described it as 'bloody murder'; all were disgusted; most will leave the Commission within a year or two. Nowhere is the futility and social danger of the 'economic' approach so clearly demonstrable as in the current state of forestry. But at least these men are awake now, and know how their community is being destroyed. They were ripe for tackling. There are others, everywhere, every day, suffering the same needless indignities, their precious social resources squandered, in lay-offs, takeovers, inflation squeezes. Bring home to them the real, basic reasons for these troubles; a local illustration is worth any amount of blackboard diagrams, worth far more than simple unsupplemented Figures 1 and 2. What an opportunity was wasted at the Cupar sugar beet showdown! Be ready for the next such event in *your* locality.

Those teaching in schools have the greatest opportunities to do good. Working often in squalid buildings with disgraceful equipment, assailed by cheapjack values and battling against the sullen hostility of pupils old enough to know they are being cheated – what can these teachers offer? Certainly little that their Training Colleges told them to offer. Long before 'O' level the children have seen through that gas. There is plenty of energy left in those youngsters who have not yet hived off into their own apathies: must this energy be allowed to blaze into aimless violence, into breaking and beating up, or when vaguely directed, into tearing down all those hated bourgeois values? Was it MacDiarmid who once complained bitterly that, although Scottish conditions engendered the greatest violence, all of this valuable violence was misapplied? But no, you certainly should not try and direct the violence against the central government's deceits, the industrialist's feudalism. You should try and seize that energy before it dissipates itself to such violence, and channel its force into building up a young Scottish community. Show them that they have a Scotland and that they will be – are, in fact – its community. This is not an easy task under the present rain of propaganda, but no more difficult, and much more rewarding, than trying to persuade them they mean something to a London-directed society, to a world so obviously bent on exploiting them that only a spray gun, a bicycle chain, or later, a damaging strike, can afford a satisfying retort. These children are your allies, they are the future.

The enemies of the future can offer them only palliatives, such as Outward Bound courses – activity in a vacuum. Scouting, now so patently tame and contrived (and politically suspect), has to hire a postmark to insist that, really, it is 'Adventure'; service youth clubs – we need say nothing on those traps that is not implicit in preceding pages. You can offer children something that none of these outmoded substitutes dare think of; you can offer comradeship in a land their own and the chance to make it the best they can – and their own part to play in the world. In this participation every school subject has a vital contribution to make. Those who teach in schools can play the greatest part in building Scotland's future.

As for the universities, we have discussed them elsewhere. The Bolsheviks, Nazis and West Pakistanis eliminated the intellectuals among their opponents by crude murder; it is to the credit of the more experienced London government that it prefers to accomplish this essential preliminary to genocide by a more effective method. Those who doubt the efficiency of such treatment may visit our universities and search for the intellectual leaders of the Scottish community. They exist, but silently, working in beleaguered groups. Interest in Scottish affairs is judged at best a time-wasting crankiness, more likely a sign of political (and hence academic) unreliability; the upper echelons are, by natural selection, fervidly Unionist, or cosmopolite, or . . . they keep quiet. These last do not need our help or our advice.

What about the students? In many faculties in Scotland, Scottish students, as we have seen, are in a clear minority. Scottish affairs are dismissed by English student politicians as trivial. These ardent demagogues are conveniently ingenuous. They are themselves led carefully by the nose away from any relevant political action; they are given noisy but harmless toys to play with – safely faraway questions like Greece, South Africa, Jews in Russia. They can bang and shout away outside the embassies of choice while the real enemies of their generation, who are destroying their future right beside them – the industrialists, politicians and economist hangers-on – smile indulgently. If there was no Greek Junta it would have to be invented. If students ever turned their attention to local affairs, the clamps would go on. And in Scotland the task is simple and obvious; we have spent much of this book depicting it. A worthy, honourable task, highly relevant and certainly exciting. A task the rootless, idealistically undernourished youth of England, say, or the United States of America would envy – and would support, could they be shown how necessary it was for all the world.

Yet drearily the agitators parade the same old placards handed out by knowing governments. Predictably they try, like their elders, to stamp out 'minority movements'; seeking independence, they depend heavily on the authorities; raging against reaction, they faithfully oppose any progress. By the familiar method of minority majorities at a 'special' meeting, they dissolved the Scottish Union of Students into the National (English) Union of Students Region No. 10; a protest lodged by Paisley Technical College – one of the rare bands of Scottish students still existing – forced a delay, but the votes of the vast numbers of English undergraduates in Scotland, while not reaching the two-thirds majority needed constitutionally to dissolve the SUS, was sufficiently powerful for the central student bureaucrats to announce that they would 'go ahead' as if the SUS was Region No. 10; and that is the outcome of the generations working at St Andrews, Aberdeen, Glasgow and Edinburgh. So this student central government loyally enacts within its own microcosm the tactics (perhaps it would, in another country, label them 'fascist techniques') of the Whitehall its members aspire to. But a rump SUS still refuses to be dissolved on such grounds and so provides an excellent growth point; the duty of every Scottish student to support that body is quite clear. Yet the task is primarily a guerilla one at first – activist encouraging activist, group emulating group, taking advantage of the clumsy and dogma-ridden enemy. Official repression whether by government authorities or their unwitting watchdogs, the tamely baying packs of the 'New Left', is helpless against guerilla tactics. The 'clamp' only grips large inert bodies. Most Scottish students detest cant, whether from careerist academics or bourgeois puppy-dogs; they want, once again, to do some solid work for what is recognisably their own community.

When sufficient force has been generated, the Treaty of Union will have to be re-examined and modified, perhaps drastically, in the light of modern conditions. Entry to the EEC by England may, anyway, amount to *de facto* annulment by England. The tide of opinion has obviously turned; though each wave falls back, the sea advances. We shall succeed – *as long as we retain our identity as a Scottish community*, and this means work *now* and for everyone. It must be continuous work, even if as simple as remembering your correct address and writing it, or knowing your correct flag and flying it.

This book does not deal with the politics of how a greater degree of independence for the Scottish community is to be brought about. Its concern is the total environment. Its aim is to indicate how independence is an essential ingredient of the environment and how it is essential for the health and prosperity of that community, and to outline ways of putting the message across and of overcoming the central government's aggression. This book contains no manifesto for any political party; indeed the weaknesses of all have been made apparent.

But surely our 'achieve the positive, resist the negative' approach must also apply to our voting behaviour? How should we vote?

We have seen how the London-based parties are quite against any practical form of self-government for Scotland and understandably so. Their election promises invariably contain home-rule phrases which in reality dissolve into time-serving and time-wasting committees or councils. The Liberals promise more, but as they stand no chance whatever of forming a Westminster government, they can afford to promise more; Home Rule is yet one more of their pie-in-the-sky escapist phrases. They show no signs of having examined the practical consequences of either the retention or the repeal of the present Union.

Of the Scottish parties only the Scottish National Party appears at all seaworthy; it has in fact been at sea for a good time but has proved surprisingly unsinkable. One has reason to believe its leaders are becoming better sailors. Although the gesturing freaks have been largely left behind, it appears to base its voting appeal on a lukewarm kind of emotional nationalism, chilled further by current economic fallacy. Hardly inspiring, as its hungry but hesitant customers testify. It has tried to tackle the Unionist parties on their own pitch and completely neglected its home ground, the ground where they fear to tread, where *they* are at a hopeless disadvantage – the spirit and resources of the Scottish community, and the bright future awaiting that community as long as it can fight free of the rising southern flood. It says much for the spirit of the Scottish people that they have voted so heavily for such an uninspiring, provincial message as the hapless SNP provides. Provincial, in truth. Dull, honest leaders; 'Our Winnie', 'Winnie's Covenant' . . . a damning seediness over all, too often not national, but nationalist.

Yet they work hard, they mean well and they are a vast improvement upon their unhappily exhibitionist predecessors. Thanks to their efforts the central government has recognised Scottish feeling; but such recognition has taken the form of throwing the whining hound a few scraps. Because of a modest success in its (futile) frontal attacks on the entrenched Centralist parties, effective Scottish political opposition has been polarised, and, since it was absurdly easy to penetrate the SNP, easily controlled. We have already described at length the contemporary more soothing, governmental approach with the humane killer in one hand.

But there is no alternative. We cannot back away. We must go on, and ensure that this 'Scottish feeling' grows and grows until it is too big for any killer, humane or not. To stop now would be disastrous. So we are obliged here to recommend a vote for the SNP, or even, a participation in its inner renewal. And if London Labour tells us the SNP is right-wing and London Tory tells us it is left-wing, let that comfort us; for the SNP is, perhaps fortunately, not a *governing* party – but a party, the *only* party, pledged to work for self-government; and as such it may be expected to, in fact should, contain all elements from Right to Left within it. Despite what its enemies say, and the bigger of its own fools also, it has nothing to do with Left or Right; it is there to ensure we can have a real Left and Right. It cannot be successfully attacked by the London parties except when it comes down to snuffle in their sops, to accept their economic fallacies, and how easily the dull earnest creature *is* lured down! But it is learning, the hard way. And as its support grows there will be many more 're-spectable' public figures working with it. (A surprising number do so already, incognito.) It must only improve with time, as the London-based parties must decay; for no able man likes being a mere pavement entertainer, and if the SNP were even a little more acceptable and could guarantee a few more votes, plenty of contemporary Scottish lights of Labour and Tory alike would move across.

When, by whatever means, the Union is modified or repealed and there is a Scottish Parliament, then is the time for picking your correct colour of man to represent you. But until then, only a vote for the SNP will be of any use at all; for only that vote is a *vote for the Scottish community*. As a corollary, one should support the SNP (which of course has no regular income) by donations; even, if one feels strongly enough, by joining, until independence, at any rate.

The London parties' hatred and fear of the SNP is a

measure of its value to our community at the present time. Their condemnation – which can be read in any newspaper at the approach of a by-election – is revealingly hysterical (reminiscent of the song-and-dance against the independence parties of the Colonies in the nineteen-thirties). Equally revelatory was the behaviour at the last General Election of a few of the worst newspapers.

No further comment is required. The medicine may not be to everyone's taste, but at the moment nothing else will save Scotland. Vote SNP, so that you can have a chance of voting effectively for your community either left or right or centre in that community's future Scottish parliament. And that is as far as we wish to influence your political 'allegiance' in this book. For, as should by now be clear, any vote for a London based party is a *wasted vote* and can only bring about an unnecessary and therefore unforgivable impoverishment of our children's future.

Remember, never engage the enemy on the ground he has chosen (that is why he chose it), never fight him with *his* weapons, he knows them as well as you know *yours*. He will seek to divide you, to refer you to a London choice, a London norm. He dare not meet you on your own ground; he dare not risk admitting you *have* a ground.

That is the dilemma of the London parties. If they decry the value of everything Scottish, they have thereby to admit its existence. If they ignore its existence, it will develop unattacked. Economically at the moment, we are on the defensive. Psychologically, the vital aspect, *they* are. We can attack them psychologically far more effectively than they are able to reply. Once we begin, their only retaliation is by means of numerical aggression and industrial colonialism; and these, as we have seen, can be defeated by community spirit. Once we begin to rebuild the community spirit in Scotland, the central government has lost. Recent elections have shown that this spirit is there waiting, even anxious to be rebuilt. Each one of us must do his part in act, in thought, in speech or in everyday conversation, to rebuild it. Whatever the insufficiencies of this book it surely supplies enough growing points to act as guide to thought and action; from the strategy outlined in it, any intelligent person can develop his own tactics in the fight. It is a guerilla war against an over-committed cumbersome enemy who can still do much damage but who, if we weaken his propaganda, can never win and must eventually pull out. Home rule votes and politicians are secondary; they come intermittently at elections. *Now* is the time for each one of us to act in his own locality, his own medium; though the future rewards are greater, we experience rewards even now in every way we strengthen our community. Every discussion group we set up, every new awareness that dawns, every admission that 'I never thought of looking at it like that before', is a very real satisfaction. We are not paying indulgences to some Whitehall hierarchy; or bowing down before some Marxist philosopher's stone. We are ensuring the continuance and flowering of our own inheritance – that strikes home to us everywhere in our hills and rivers, our cities, our buildings, our songs, tales, our parents and our children; we are fighting for what has always been accorded the noblest thing a man can fight for – the defence of our own community. And this fight deals no death, harms no other people; it helps others. We are fortunate that at a period when to the delight of the East, much of the West has so confused itself and is facing moral and economic dissolution, we in Scotland have such a clear, noble and rewarding task before us – before each one of us; starting *now*.

PART IV

1. Constraints on Expansion

It may now be clear that, among the conflicting forces at work upon a community, the greatest is the unrestrained urge for economic expansion. We shall call it the industrial imperative. Yet the erosion of choice, the decline in morality, the destruction of our psychological and social resources which it causes is effected so insidiously, that it is hard to find any one point at which one may logically cry out 'enough'. A rearguard action in one front merely means the neglect of another, where the erosion may proceed the faster. A sense of despair envelops the thinking man. Cublington is saved at the expense of Foulness and Wigtownshire. As the Conservation Society remarks: 'Even those who do look a little farther ahead are usually content to excuse their inaction by accepting the claims of the technological optimists. . . .' When, as in 1971, there is massive and growing unemployment, there is no obvious and immediate solution to the problems of gaining a livelihood except to generate an expansion of the economy. All political parties, every Trade Unionist, each small business man sees it as a device for his own immediate salvation. It is a personal event, and though the concomitant disadvantages are becoming more obvious, and are being voiced more cogently, practically every voter votes with himself in mind, and votes for expansion. Live now, pay later.

How then, in the light of this inherent urge within man, can the authors hope to carry their message into the realm of practical politics? People may agree with us, but the industrial imperative will remain, each one of us aiding it, as each one of us is aided by some small part of it – the man out of work who gains fresh employment, the broker on a rising stock market, the politician who is able to face his voting public? The purpose of this chapter is to reveal how close man is to the end of his term of wanton expansion, not because he wills it to cease, but because globally he is living beyond his means and out of balance. Once this point is understood and accepted many of the shibboleths of economic expansion collapse like a card house in a draught of clean air.

There are some 3·6 billion people on the earth and during the days from April 13 to 17, 1970, it is probable that more of them shared a common bond than at any time before or since. The American space-craft Apollo 13, with three men on board, suffered a technical fault which gravely imperilled their survival or return to earth. Modern communication media enabled us all to share their experience and for four days the thoughts of the whole world were centred on them. Why? It was not because technology or even American technology had failed. Even the Russians wanted them to survive. It was not because the deaths of three men in a day was not a commonplace experience. As many die any hour in road accidents. It was because out there we could feel the human spirit coping with a situation in which one false move would spell disaster and not instant merciful disaster, but a disappearance into space without recall. If the actions of Mission Control had been unsuccessful these men would have orbited in space until their supplies ran out and then they would have slowly died, probably of oxygen lack. Their need would have outrun their resources. Television would have provided each one of us with a death bed perch.

Look now at the technical side of the event. A space capsule is a world of its own, furnished with energy, food, air and water. The quality of the air and water is assured by recycling certain things, but to do this consumes energy. Some of this energy comes from generators that react to the sun's rays. Through these energy sources, in this case oxygen, lay the whole means of survival for the inhabitants. On that fateful flight, something smashed the oxygen tanks, reducing the energy supply of the space capsule to a third of normal. A capsule on its way round the moon carries very little spare. Suddenly the men, who till then had lived in a technological paradise where everything they wished to do could be done, from eating, drinking, monitoring the world with a television camera and generally fussing away at little but interesting routines, were faced with the other extreme. Every resource in the capsule had to be conserved. Movement created heat and sweat that consumed energy in de-humidifying the air and cooling it. Breathing was anti-social for it consumed oxygen and that had to be regenerated. It was survival by inaction. To the boredom was added the unresolved fate ahead of them.

If Apollo 13 had not suffered damage, but it had contained through imprudence or accident twice as many men, the result would have been the same. The ratio of resources to men would have been too low, consumption too high. That would have been a fatuous mistake on the part of the Space Mission Centre at Houston.

The world is also a space capsule. Larger, more diverse, less sophisticated technically, but finite all the same. Because men have been into space, and we have all seen pictures of our world from a distance of 250,000 miles, it is easy to visualise the earth as a capsule in space. In those distant pictures taken by the astronauts we see a globe. We know it is ours; made of minerals, rock, gas and water. We know it is surrounded by an atmosphere. Superimposed on this globe is life, a complex inter-relation of men, animals and lower creatures. We can roughly estimate the numbers of mankind, but to put an accurate figure on other species is beyond our ability. All that may be said is that they outnumber man by a factor of many millions. Every one of these creatures from the smallest virus to the largest mammal consumes energy. Some of that energy is implanted in the earth in forms like coal and oil, stored up during the epoch when man did not exist, when the earth absorbed energy from the sun faster than the creatures upon it consumed it. That era is over. Man is now consuming the stored energy faster than it can be replaced. He is dipping into a larder that he has no means of replenishing. However, a daily ration of energy still reaches the earth – sunlight. It is through this sun energy that the world built up reserves in the past, and it is by this energy that every creature lives today, for it is by photosynthesis that all life continues. The amount of this bounty, this income freely radiated at us, is the astronomical one of between 1×10^{17} and 2×10^{17} calories per year.[27] More reaches the tropics than the poles. Clouds and atmospheric pollution reduce the income.

This energy is metabolised into some sort of living organism. Animals consume plants. Man consumes both. He is at the apex of the food-consumption pyramid. Since about one-sixth of this energy reaches a form useable by man, it is a simple matter to calculate how many men could live on the planet earth. The answer is about thirty billion. This calculation assumes that no carnivores are left to compete and that no pests or crop diseases still occur, an unlikely situation. From past experience we know and understand that when such a situation exists there is created a perilous biological imbalance. There is ample evidence, even today, to show us that when a population lives at the edge of its resources, it takes but a small reverse to decimate it.

According to the National Academy of Sciences (NAS) Man and Resources Committee, thirty billion would represent 'chronic near starvation for the great majority with massive migration to less densely populated lands'.[28] The diet would be miserable, a dreary factory-farm world, an utterly pointless existence. At the present rate of population increase it would take till 2075 to reach this crisis level. Man has a century at most to solve his problem. Population is therefore a global problem.

However, our concern at the moment is not with population in total, but in relation to resources and consumption, as highlighted by Apollo 13. The maximum figure for world

CONSTRAINTS ON EXPANSION 123

population of thirty billion represents an average of 520 persons to the square mile of the earth's surface. It does not seem a large figure and it is already exceeded in certain countries. For example, today Japan has 693 persons per square mile, Holland 1,045, England 908, Scotland 171. It is important to visualise the effect upon industry as the world approaches the end point of her complement of humans. Since this point is reflected in the total orientation of every resource towards making food, it follows that those countries who have a food surplus or potential for surplus will have been under pressure to accept immigrants. With food in short supply throughout the world, countries with surpluses may store against bad harvest. They are unlikely to be exporting it. Their populations cannot be fully employed producing food, but their viability in this connection will have given them a distinct advantage in industrialisation. They will have been able to import the highest calibre of immigrants. They are unlikely now to be significant exporters of food or importers of industrial goods. The over-populated lands will be the first to suffer. Since, as likely as not, they will be over-industrialised, they will be putting out more goods into an ever diminishing market which men will progressively lack the means of buying. They may find they have made the bargain of Faust.

Our task – indeed, as intelligent human beings, our obligation – is to look at the position globally, to identify and separate the parameters of survival and then to measure how close we are to danger. The parameters are easily distinguished simply by thinking in terms of the space capsule. Some are obvious, such as numbers of people, resources, energy, food and consumption. Their interaction adds up to the standard of living we may or may not enjoy, or *in extremis*, whether or not mankind as a race can survive. The complexities of these matters have been superbly presented by the Ehrlichs in their 'Population, Resources and Environment', and in less analytical detail but more globally by the Man and Resources Committee of the NAS, who succinctly summarised their conclusion:

'Indeed it is our judgement that a human population less than the present one would offer the best hope for comfortable living of our descendants, long duration of the species (man) and preservation of environmental quality.'

Indeed one may readily find scientists[29] who will argue that a population of one billion, a quarter of the present, is about right for this earth. These conclusions come from an analysis of the interaction of the obvious parameters, already mentioned. Others, less obvious, will emerge as the discussion develops.

Population

If thirty billion people represents the brink of survival in this world for all, the present population of some 3·6 billion seems conveniently far from this terrible situation. Table 6 illustrates how head-in-the-sand this attitude is – this truly pressing nature of the time scale.

Table 6

Source: Kormandy, *Principles of Ecology*

It took:

75 years from 1850 to 1925 for the world to increase from 1 to 2 billion
37 years from 1925 to 1962 for the world to increase from 2 to 3 billion
15 years from 1962 to 1977 for the world to increase from 3 to 4 billion

and he estimates that it will take:

10 years from 1977 to 1987 for the world to increase from 4 to 5 billion
7 years from 1987 to 1995 for the world to increase from 5 to 6 billion
5 years from 1995 to 2000 for the world to increase from 6 to 7 billion

We may as well face the fact that as a species we have already over-produced. Moreover, no matter what is done to spread measures of natural population control, the population of the world will grow to around seven billion by the year 2000. This is partly because the under-developed countries have a 'much greater proportion of people in their pre-productive years, the size of the child-bearing fraction of the population will increase automatically . . . even if great progress were made in immediately reducing the number of births per female in those countries it would be some thirty years before such birth control could significantly slow down population growth.'[30]

And it is partly because, even in highly developed countries many people favour large families. Only Bulgaria and Japan have achieved population stability. With us only the city of Aberdeen has done so.

We can draw little comfort from the hope that the population projections are exaggerated. Again and again in the immediate past, even well advised bodies such as the UN have underestimated population growth. Curiously, this growth has its roots not in the reproductive urge of mankind so much as in the fruits of his technology, which diminished infant mortality and allowed greater age before death, and which permitted unrestrained access to the world's energy store, today being used many times faster than it is being replaced. The sombre conclusion of the NAS committee was that:

> populations may level off not far from ten billion people by about 2050 AD – and that is too close, if not above the maximum that an intensively managed world might hope to support with some degree of comfort and individual choice. . . .'

However, might not mankind do what he has done so often already? Emigrate, this time to another planet. Patrick Moore assures is that it is inconceivable that there are not other inhabited planets in the universe, though none, he declares, within our solar system. It is a simple calculation to show that it is impracticable to export billions of people from a world living on the edge of its resources to distant spots, light years away.

Ehrlich,[31] with ineluctable arithmetic, calculates that to export 3 days of current world population growth would take an effort costing the annual GNP of the USA and that even if Venus, Mars and Mercury could sustain life, it would take but another fifty years to fill these planets as well. Mankind can put off the solution, but the solution must in the end be found.

Resources

We have seen that we shall never live to become so crowded that there will not be room to move about. At the worst there may be many more Tokyos and Londons, without the historical attraction of either. It is food through sunlight that will limit that population. But what about production? The world economy can expand on two bases: an expansion to cope with the increase in world population at present living standards, and another, higher rate due to 'growth' and rise of material living standards. How do the world's physical resources match the demand for them? It is a common assumption, even by economists, to assume that there are ample resources; that as one deposit is worked out another will be found. Such complacency has its justification. In the thirty years up to 1969 discoveries of oil reserves have been at such a rate that the world's proved reserves have risen rather than fallen in terms of years of stock in spite of accelerating consumption. There is a single mountain in Brazil and another in Australia that could feed all the world's steel mills for a decade. There are ore deposits in Australia and Greenland just waiting a rise in price to justify economic exploitation. Surely, therefore, it is permissible to visualise a steady expansion of resource consumption. The demand is undoubtedly there. Industry will see it is added to. But the world is not infinite. Resources, if not replaced, cannot last for ever. Crudely, the question is whether we need worry about it now or will mankind's inherited resources serve several more generations or even centuries. Briefly stated, non-renewable mineral resources are in such short supply that within thirty years economic expansion as we know it now will be impossible.[32] An economist ironically remarked that 700 million poor Chinese presented the world with a grave problem, but nothing like the problem that would be created by 700 million rich Chinese, for look how much more they would consume.

In the past thirty years man has consumed as much mineral wealth as in all time past. Since it is only natural that the present 3,000 million economically less fortunate individuals in the world will aim at (but cannot attain) a North American standard of living, consumption will rise at an even steeper rate. To reach US material standards, according to Ehrlich, would require the extraction of a further 30 billion tons of iron, 500 million tons of copper and lead, 300 million tons of zinc – that is to say about seventy-five times the present annual rate of production. There is no evidence to suggest that reserves on such a scale exist or may be found or that the world's eco-system could sustain the pollution that would be created. We are told by the NAS[28] that:

> 'true shortages (which) . . . exist or threaten for many substances that are considered essential for current industrial society; mercury, tin, tungsten and helium for example. . . . It will only take another fifty years or so to use up the great bulk of the world's initial supply of recoverable petroleum and natural gas. . . .'

Alexander King, the Scientific Director of OECD made the point in a homely way at the 1970 annual Science of Science Foundation lecture in London; 'If everyone in the world had access to one newspaper a day and used two sheets of toilet tissue, the world's forests would be exhausted in thirty years.'

The optimist will suggest we turn to the poorer minerals, to sea water, to re-using spent materials. But these operations are extremely expensive in energy and thus make for expensive products. Moreover, the amount of material to be processed for a given ton of pure product is so enormous that it automatically creates vast pollution problems. For instance, to extract one ton of zinc from sea water would require the processing of twenty-two billion gallons of sea water. The process itself may require the consumption of fresh water far exceeding the supply. Technological optimism on these lines merely obscures reality.

'During the next century adequate supplies and equitable distribution will not be achieved merely by recycling scrap metal nor by processing dozens of cubic kilometres of common rock to supply metal needs of each major industrial nation. When the time comes for living in a society dependent on scrap for high grade metal and on common rocks for commercial ore, the affluent society will be much overworked to maintain a standard of living equal to that of a century ago. Only our best efforts in all phases of resource management and population control can defer that day. . . . Is it possible that American standards of consumption are excessive, and to accept them as world targets is a tragic mistake?'[38]

Technology has achieved such a remarkable level of development that ill-informed optimism amongst the scientifically ignorant is understandable. After all, man can now transmute the elements – the alchemist's dream – grow protein artificially or make aluminium out of common rock. But there are several essential ingredients to these hopeful recipes, some of which are limiting factors. Take energy, for example. It takes more than eggs to make an omelet. Energy, in the form of heat, is also needed.

M. King Hubert of the US Geological Survey has reckoned that if fossil fuels continue to be used as the major source of energy in the world, oil and shales might last a century, coals a hundred or two years more. Neither solar energy nor tidal power nor continental drift can ever hope to offer more than a few per cent of even our current needs. There is a widespread view that nuclear breeder reactors will solve our energy

CONSTRAINTS ON EXPANSION

problems, but already the insufficiently considered problems of nuclear waste and leaks are creating alarm in large countries like the USA. It is true that if the fusion reaction were made to work, the world would have by present standards 'unlimited' energy, but this has yet to be achieved, or even hoped for, at the present level of either the science or technology of the subject. And there is still the radio-active by-product to consider, even though it is expected to be less. Radio-active waste disposal is an unresolved problem. After all, if irradiation is now a practical means of eliminating unwelcome insect life, might not an imprudent gesture or reactor accident create similar hazards for men? The generation of energy itself uses even more energy. A fundamental natural law, known as the Second Law of Thermodynamics, states this. No contradiction has ever been found. From it we know that every attempt to generate, say, heat or electricity calls for the dissipation of part of the original energy. Mankind's every gesture of self-indulgence automatically creates a further run down of the planet. Scientifically expressed, global entropy is increasing.

A vital resource, reckoned of great account elsewhere but very little considered in the moist British Isles, is water. It is not just a domestic commodity to be drunk, nor yet simply an industrial solvent, widespread and considerable though these uses are. It is also used in vast quantities for cooling. The amount of water available could conceivably limit industrial development. England depends on water from Wales, the United States of America on supplies from Canada. Water supply is a matter of habitual crisis in the north-east of Brazil. In continental United States of America it has been estimated that within the decade 60 per cent of the available cold water may be used for power station cooling. Hot water effluent creates considerable biological problems, not to be dispelled by such incurable optimism as expressed recently in *The Economist*, that: 'efforts are also being made to develop methods of dispersing the heat so quickly that it does not raise the temperature.'[34]

In Europe the need for water has been met more and more by pumping from underground. Borgstrom[85] estimates that '... man is removing water from the continents faster than the hydrologic cycle replaces it; that the people extract three times what the cycle returns to accessible reserves'. Water capital is being depleted at a shocking rate. In England, where in many regions water is subject to continual recycle, the toilets at Staines literally provide the drinking water at Chelsea.

We come to the decided conclusion that man and his resources cannot be separated from the numbers of men. His major resource is land. Enke observes,[56] 'Twice as much labour and capital will not double output if there is a scarcity of equally useful land'. One has only to compare two countries with very modest natural resources, but vastly different population densities, to see the interaction. India with 395 persons to the square mile has one of the world's lowest standards of living, while Iceland, a stern island in the North Atlantic, with three persons to the square mile, has one of the world's highest incomes per head. Both are culturally highly developed, India amongst an élite, Iceland amongst everyone.

The world population density is an average of fifty-four persons to the square mile. This forms at least some basis for comparison of latent wealth. Of course some areas lack vital resources. Others are over-burdened with people. Large sections of Northern Canada are called the barren lands. Much of South Western United States of America is completely arid. Half of Brazil lies in Amazonia, a region that can never give a rich agriculture, even if its other problems were to be solved. The Green Hell may become a green paradise, but never a tropical bread basket. The Highlands of Scotland have an acid soil and too much peat, but its coastal and water resources are superb. England is so over-populated as to be unable to feed 40 per cent of her population. The criterion of potential richness is the sum total of population density added to other resources; oceanic, coastal, pastoral, mineral, hydrologic, climatic, cultural, social and psychological. Scotland, for

Table 1

Gross National Product and Population and Area

Source: UN Statistical Year-Books

COUNTRY	AREA Sq. Miles	1958 GNP $/head	1958 Population density persons/mi²	1967 GNP $/head	1967 Population density persons/mi²	1958–67 % increase GNP	1958–67 % increase Pop.	Total pop. 1968 millions	GNP per head/pop. density 1968	Increase GNP $ per head 1958 to 1967
Australia	3,037,000	1,410	3·4	2,210	3·8	56	11	12·031	581	800
Austria	32,370	750	216	1,470	221	96	2·3	7·350	6·6	720
Belgium	11,781	1,160	768	2,170	775	87	0·9	9·619	2·8	1,010
Brazil	3,286,000	150	20·3	320	26	113	26	85·12	12·3	170
Canada	3,851,000	1,990	4·4	3,000	5·4	50	21	20·77	555	1,010
Czechoslovakia	49,000	1,540*	274	2,260	285	46	4	14·36	7·9	720
Denmark	16,600	1,100	271	2,320	286	109	5·5	4·87	8·1	1,220
Finland	130,000	925	33	1,520	35	64	6·0	4·68	43	595
France	211,000	1,100	212	2,550	233	131	9·9	49·92	9·4	1,450
Germany	93,300	1,060	558	2,290	616	116	10·3	58·01	3·7	1,230
Ghana	92,000	120	70	210	91	75	22	8·37	2·3	90
Iceland	39,700	2,370*	4	2,100†	4·9	−11	22	0·20	428	−270
India	1,260,000	low	—	low	412	—	—	523·8	—	—
Netherlands	12,900	850	867	1,990	968	134	12	12·743	2·1	1,140
New Zealand	103,000	1,400	22	1,645	26	18	18	2·751	63	245
Norway	125,000	1,140	28	2,550	30	123	7·1	3·819	118	2,210
Scotland	30,411	≃1,100	171	≃1,500	171	40	0	5·187	8·8	400
Sweden	173,000	1,510	43	3,230	45	114	4·6	7·918	72	1,720
Switzerland	15,900	1,400	320	2,780	380	99	18·0	6·147	7·3	1,380
USSR	8,650,000	155*	23	1,130*	27	—	17	237·798	—	—
England	58,300	1,300	776	1,720	862	32	11	50·100	2·0	420
USA	3,615,000	2,610	48	4,380	55	68	14·5	201·152	80	1,770

* Computation uncertain due to parallel exchange rates.   † Since made a magnificent recovery.

example, so frequently presented as a poor country, with 171 persons to the square mile and a total population of five million would seem by any measurement a remarkably well-endowed nation in relation to her population – and that is the point. Examine second last column, Table 7.

Consumption

Since the finite nature and diminishing state of the world's non-renewable reserves have been established beyond doubt, the relation between numbers of men and their consumption is obvious. Thus in the long run, in order to maintain consumption man should cut his numbers. Ethically this is an impossibility. Mankind has no recourse but to accept reduced consumption. The concept of increased volume output per person as a route to increased productivity and income is as incurably optimistic as the dreams of the alchemist. Lead can be transmuted to gold, but the price is beyond our means. With insufficient resources to go round, mankind is in the position of a party of explorers in the High Arctic, who, trapped by the winter, must assess their food supplies and plan their consumption with due care for the long winter ahead, sharing them equally. Man dare not consume his supplies blindly until he sees the last few jars on the larder shelf. However, before we discuss the implications of this inevitable state of affairs, there is one other limiting factor to consider, less obvious and perhaps more potent than all others which will limit the expansion of industry. It is our environment.

Environment

The dynamics of yeast cell division in a sugar solution reflect our man-made society. The cells increase in number slowly at first, and then rapidly, the stage the world is in now, and finally the dynamism is lost as the concentration of alcohol formed rises and the yeast cells choke in their own by-product. Finally the system dies. Someone else drinks the resulting wine.

Man cannot live without producing excrement. It applies

to his domestic, his agricultural and his industrial activities; even, in rich countries, to his social activities. For a very long time he has simply discarded the unwanted by-products of his life into the rivers, oceans and air, giving no more thought to the matter. The blighted English Midlands, the Ruhr, Galveston Bay and the Great Lakes bear evidence of this crude approach. Even so, until very recently such action was considered the necessary price of progress, having little detrimental effect than that of creating a bleak environment. Now it is known that pollution can have a deadly end effect. It can block biological chains, upon which we humans depend for our survival. It can alter the earth's heat balance, even perhaps cutting down the incoming source of life, solar radiation. It can create gene mutants, and change whole populations. It can even end human life on earth.

For example, there is now unshakeable proof of the ill effects of DDT. This miracle insecticide solved many of man's problems when it was first introduced in the late 1940s. What was not appreciated then was that this chlorinated hydrocarbon is non-degradable. Carried by a sequence of biological chains to become selectively concentrated in the fatty tissues of birds and mammals it there causes diseases like cerebral haemorrhage, hypertension, portal cirrhosis of the liver. In animals many species cease to reproduce. Ehrlich has computed that if DDT use was stopped now it would still be another thirty years before man knew the worst likely effects of this chemical. During that time the concentration in the sea will rise, with horrific consequences as yet barely imagined. No one has yet traced the source of the mercury poisoning that is killing the American West Coast Tuna industry.

Man is only just learning that he cannot go around changing the world on a grand scale for his own convenience. The celebrated Aswan dam created more problems than it solved. During the time it took to build it, more Egyptians were born than the newly created fields irrigated by the dam could feed. The vast new area of stagnant water promoted the serious parasitic disease of bilharzia. Los Angeles was once a desirable residential area, where everyone owned a car. Now pollution from cars is such that forests 200 miles distant are dying from the fumes and in windless weather children are forbidden to exercise so as to cut down the risk of bronchial disease.

What is only now being clearly appreciated, and even then only in principle rather than detail, is that man is by no means a separate part of the world, but part of biology. All organisms, including man, are linked in a dynamic equilibrium. The death of one undesirable insect may simply remove the predators, which enables another to gain a worse hold. Many pesticide projects are now seen as 'tragic blunders'.

In spite of being *homo sapiens*, man is in danger simply because he is at the apex of the food pyramid. Ehrlich explains the complex but vital scientific argument:

> 'Because of the loss of energy at each transfer along the food chains, the higher a position the population occupies in that food chain, the smaller the population will be... this means that if a poison were applied that would kill most of the predators and herbivores in an area, it would be more likely to exterminate the population of predators than the population of herbivores. Purely by chance, members of the larger population would survive. It would not be necessary to kill all the individuals of any one species to force it to extinction.'

Man is the highest predator, and therefore the most vulnerable.

Economic growth
Economic growth seems to be the catchword no politician and few economists will deny themselves. The economist Mishan remarks:

> 'The public has a remarkable faith in the ultimate beneficence of industrial progress, a faith which it seems to think,

could in the last resort, be redeemed by modern economics. Contrary to their fashionable phrases about the need to face change, those who proclaim themselves to be in the vanguard of new thought prove to be in the iron clutch of economic dogma, much of it provided by famous economists of the past as a guide to policy in a world different from our own. Free trade, free competition, sustained economic growth, the free movement of peoples – these were, for Britain at least, the dominant economic aspirations of the nineteenth century. . . . Stated more positively, the younger generation will be facing the future with honesty only when it brings itself to face the strain of thinking through the consequences, tangible and intangible, certain and speculative, of the current drift into the future and, in doing so, recognises that in the new world the old liberal economic harmonies are not to be found; that on many issues painful choices have to be made, and that in some cases the needs of men and the needs of technology may prove to be irreconcilable'.

The issue to be made plain is not whether we should have economic growth, nor what to do with the gains, if any, resulting from it, but how long mankind can fool itself into believing growth is an open-ended situation. Herman Kahn, in his *Year 2000*, devolves much of the discussion around the predictions of economic growth. Though published in 1967, it is astonishing how naive it seems even only four years later. To do Kahn justice, he discusses several 'alternative nightmares', and at least one seems to bear a remarkable resemblance to present-day events.

If at any point in time it is accepted that economic growth, in quantity anyway, cannot persist indefinitely, whatever the limiting factors of pollution, resources or population, then mankind has to face a new form of life. The problems posed by any cessation of growth are enormous.

Firstly, vast segments of the human race have been conditioned by their leaders into thinking that an annual rise in their income is as inevitable as that day follows night. The developed countries, particularly, adopt a consumer-oriented attitude to production. Advertising is a major feature of everyday life. The people are not going to take kindly to being told by their leaders that this has been a false situation. The frantic wrigglings of Trades Union leaders and of politicians in the United Kingdom today demonstrate the frustration ensuing the moment economic expansion declines. 'The preoccupation of the (American) liberal with production,' J. K. Galbraith remarked, 'is an interesting example of the hold of conventional wisdom in the face of changing circumstances'. Leaders of all sides demand reflation and 'growth'.

The second, even greater problem is that economic expansion has created the hire-purchase society: paying yesterday's debts with tomorrow's inflated money. Such a situation is demanded by the industrial imperative. The re-think through to a society based on economic stability is going to be hard, particularly hard for those industrial nations saddled upon a runaway industrial cart horse. The problems raised will be discussed in succeeding chapters. For the moment let us merely frame the picture.

The most striking thing is the rapidity with which the concept of economic expansion has ceased to be a logical and natural state for mankind. Even as recently as six years ago the economists Barbet and Morse could write:[37]

> 'Few components of the earth's crust, including farm land, are so specific as to defy economic replacement, or so resistant to technological advance as to be incapable of eventually yielding extractive products at constant or declining cost.'

Put otherwise, 'Someone's always got a technological fix'. We can always use leaner ores or recycle old cars to the factory. W. Rostow[38] could pronounce the gobbledy-gook that: 'A general requirement . . . is to apply quick-yielding changes in productivity to those most accessible productive resources. . . .'

CONSTRAINTS ON EXPANSION 137

K. E. Boulding described man's approaching situation rather aptly.[39] Man is still using a 'cowboy' economy; to use and waste at will. What matter that one burns a whole tree to boil a billy of water and broil a single steak. There are other trees. Man now has to face up to a 'spaceman' economy in which resources must be conserved. Yet for industrial society, as Allan Wager remarks, 'The idea of keeping the economic plumbing full with the least possible pressure and flow is still unthinkable.'[40]

The concern is rising, faster it seems, in that citadel of capitalism, the United States of America, than elsewhere. This is not simply because they have got it all, and can afford to look cynically at the rest of the human race. Rather it is because, as the world's most advanced industrial country, they can see so much more clearly the consequences of their industry. *Science*, the weekly magazine of the American Association for the Advancement of Science, contains more intelligent, informed and balanced discussion of the problems than ten European journals lumped together. Yet the United States of America is a country less densely populated than any European country and has less even than the world average. This element of spaciousness in the United States of America is important, because it is in a spacious environment that the entrepreneurial instincts of man are most tolerable and least harmful, as the Conquistadores found in South America. A car can have a violent skid on an empty road and get away with it. A gelignite bomb can explode in the ocean and hurt no one. One breathes safely because there is so much oxygen and plants are continually replenishing it. Even if the population reaches many times present levels there would still be plenty of oxygen to spare.

The concern is not simply that of the well-paid scientists. Even Henry Ford has his doubts.

> 'Modern industrial society is based on the assumption that it is both possible and desirable to go on forever providing more and more goods for more and more people. Today that

assumption is being seriously challenged. The industrial nations have come far enough down the road to affluence to recognise that more goods do not necessarily mean more happiness. They are also recognising that more goods eventually mean more junk, and that junk in the air, in the water and on the land could make the earth unfit for human habitation before we reach the twenty-first century.'[41]

The Duke of Edinburgh echoed these sentiments in London in November 1970.

Amongst liberal economists a common argument for economic expansion is that it provides the sole means of helping the under-developed countries to raise themselves from an economic level of existence that is in many ways no more advanced than in Roman times. Both Kahn and Ehrlich have shown that far from helping, many of these countries will in relative terms sink farther back. But since we are sceptical of the economists' projections, perhaps we should not draw too much from them. As Table 7 shows, the gap between the rich and the poor is widening. The President of the World Bank, Mr Robert McNamara (1970) deplored the poverty of international aid which for most rich countries runs at a paltry one per cent of the GNP.

The economist Mishan[42] has pointed out:

'If rich countries, in response to a moral challenge sought to convert themselves into an arsenal to provision the hungry areas of Asia and Africa, a case could be made for retaining economic growth as the chief goal for some considerable time (in wealthy countries).'

The NAS Committee take the view that all countries can never enjoy the North American standard of living, for there are not enough global resources to go round. It follows that if the rich do not give their wealth away, the poor will eventually attempt to acquire it by other means. Conflict seems not less likely in the materially prosperous seventies, but an inevitability.

Continued economic growth within the richer countries is bound to exacerbate the situation.

With the possible exceptions of the USSR and Canada, every major industrial nation today, including the United States of America, is a significant importer of raw materials. Super-tankers and giant cargo vessels have diminished to a tiny fraction the cost of their trans-shipment from source to industrial port in Europe, Japan or the United States of America. But as the demand for these raw materials grows, as the need for food in the food exporting countries increases, as the raw material producers seek to develop their own economies, there will follow shifts in power. Until very recently the developed nations held sway through their control of finance, and the buyers' market in raw materials. What will happen as this changes to a sellers' market was vividly demonstrated by the show of strength from the Organisation of Petroleum Exporting Countries (OPEC) at Teheran in 1971, when they squeezed £500 million more a year out of the oil companies. These Arab countries demonstrated where the real power lay – in the resources. Not the combined might of the petroleum companies, whose cash flow far exceeded that of the oil exporting countries, nor the governments of several developed western countries, like Britain, the United States of America or France had any power to move the OPEC bargainers from their ultimatum : no money, no oil. The developed countries will now cry out for international control of resources.

But international control will take time to achieve, for as a prelude the real strength, the balance or otherwise of each nation will have to be revealed by test. That strength will lie in several factors, but a significant one will be the level of its resources in comparison with the number of people it has to support and its population density. Any nation that does not have a firm grip on its resources, marine and territorial, should seek to improve the situation at once. A battle for control easily won in 1971, will be bitterly contested in the year 1991. It may take more than that time to bring the world to its senses, and in the meantime for the crowded, developed countries, the ability to gear their own economies to the realities of the future will be a measure of their future viability and hence their influence in world affairs. This is no time in history to remain a colony. This is the time for every community to take stock of its position – old nations such as Scotland must regain the title deeds of their heritage.

PART IV

2. The Balanced Communities

The previous chapter leads us to two conclusions. First, that resources being limited, some measure of control of use is necessary. We leave this discussion to the next chapter and deal now with the considerably more complex issue resulting from the conclusion that economic growth is a doomed philosophy and a practical disaster.

Whatever the determinants of society may be it is man that matters, and Lee Dice, a geneticist,[43] writing of the lessons to be learnt from 'primitive people', said:

> 'In a world in which our heads are spinning under the impact of an overload of information, studies of primitive man provide, above all else, perspective. Civilised man is a creature who each year is departing further and further from the population structure that obtained throughout most of human evolution, and that was presumably of some importance to the evolutionary process. At the same time he is not only living far beyond a reasonable energy balance, but is despoiling the resources for primary production so as to narrow increasingly the options available to redress the imbalance. The true dimensions of the dilemma that our present course has created are only now emerging. The intellectual arrogance created by our small scientific successes must be replaced by a profound humility based on new knowledge of how complex is the system of which we are part. To some of us, this realisation carries with it the need for a philosophical readjustment which has the impact of a religious conversion.'

Without economic expansion, into what alternative forms

of endeavour will the entrepreneurial urge be channelled? How is unemployment to be held back? How will politicians keep office? What will become of the determinants of society? Man needs challenge or he withers, so stagnation is unacceptable. So often, man, when faced with an all or nothing situation, will go for all; will say, to hell with the consequences. A sustained and high level of education will be needed to bring the message home. The reaction will be strong. Those political leaders who gain power by their apparent ability to create economic expansion, are likely to be very reluctant converts. There is a real risk that mankind will simply ignore the warnings and plunge on. It was possibly this risk that prompted Professor Ehrlich in 1971 to assess mankind's chances of survival into the year 2000 as about one in fifty. Such long odds do not, unfortunately, frighten people. Rather they seem so far-fetched that they can be discarded as mere haverings.

The chief alkali inspector in England, the man responsible for watching over pollution and its effects, in presenting his annual report for 1969, scorned the current concern of the ecologists. Industry sighed with relief. His Scottish counterpart has publicly stated that he abides by the old Yorkshire dictum that 'Where there's muck, there's brass' (money). Murray Gellman, the Nobel prize winner, has suggested a period of 'technological abstinence' as a means of gaining time. The world's manufacturing industries could concentrate simply on raising the standard of the under-developed countries. This is a counsel of perfection. The nations of the world have not yet learned to work together sufficiently harmoniously for such a course to work out, nor are the issues widely enough appreciated. The instinct of competition between economic groupings, the realisation that the material standard of living is in many ways not just a simple index of Gross National Product, but the degree to which one's own is ahead of others, spurs technological innovation, and economic development.

Some alternative to stagnation, therefore, must be found. The industrial workers will not accept a reduction in their material standards without a fight, no matter how illogical that fight can be shown to be. It may be possible to educate their leaders, but on past record, in the UK anyway, this will be a slow process. As recently as April 1971 the Secretary of the Scottish TUC was able to remark, without being challenged, that unemployment was directly attributable to the policies of the Conservative government. Mr Jack is not stupid, so one concludes he is either naive, ill-educated or cynically making cheap political capital at the expense of the future of his union members.

Gellman's proposal has another weakness. It is contrary to man's instinct not to push the frontiers of knowledge ever farther back. The increasing number of unemployed scientists and engineers may become almost as powerful a voice as that of the industrial workers. There seems but one solution that meets the predicament. The economy must continue to expand, but in quality, not quantity.

What does this mean? It means an end to the idea that the desirable end of business is the growth of large units and vast accessible markets on the grounds that in this way the maximum number of goods can be made for the least unit cost and sold with least restriction. Such a concept needlessly consumes resources and creates unwanted demand. In its place must come diverse innovation: diverse because ecology research has shown that through diversity is to be found stability, an essential ingredient in the harmonising of the social resources of a community, as of the whole world; and innovation because it is the true flux of industry, whereby change and upgrading of product is achieved and human ingenuity employed and satisfied. Moreover, innovation can satisfy the ecological criteria of diversity, because there is ample evidence to show that innovation is far from being the preserve of the large organisation.

It is no part of the author's purpose to usurp the economists. The task, the very real and immense task facing them is to find means of creating stability and diversity, reduced volume growth and increased qualitative growth. Since economists have for the most part been concerned with finding ways of keeping the economy expanding, and preventing depressions and business cycles, they may not take kindly to finding their researches have been aimed in the wrong direction. To brake, halt and find a new avenue may not be to their liking. However, much of the information discovered and the theories evolved will find ready application to the forthcoming state of affairs. Knowledge is seldom useless.

We content ourselves with giving a useful analogy. The present concentration of economic units is like the spraying of a powerful insecticide with a view to eradicating a pest, a serious competitor for the consumption of the growing crop. But often, not only is the offensive insect wiped out, but also their immediate predators, giving free rein to invaders. On a longer term basis the insecticide may disturb biological chains which may stunt overall growth of the desired crop. Ehrlich quotes a marvellous example of a cotton crop in Peru which suffered in exactly this way. The take-over bid is analagous to the insecticide. By reducing the diversity, economic links are broken. The take-over company can flourish so long as the economy is expanding somewhere, but it may have dealt a death blow at the local economic potential. Any subservient community can cite hundreds of examples.

Diversity is rapidly vanishing. The uniform appearances of High Streets in every town in the British Isles testifies to this inbuilt urge of economically predatory firms to destroy diversity. The solution to the problem lies in two practical steps. One is a continual recreation through native innovation. It may be likened to cell renewal as old cells wither. The other solution is political and will be dealt with in Chapter 4.

It is useful first to examine the current attitude to innovation in Europe. Harold Wilson, addressing the Council of Europe in 1966, warned:[44]

'An industrial helotry under which we in Europe produce only the conventional apparatus of modern economy, while becoming increasingly dependent on American business for the sophisticated apparatus which will call the tune in the seventies and eighties.'

Like many politicians, industrialists and science policy makers, especially in Europe, he has seized a key proposition about the industry and the technological age. 'You gotta be big to innovate.' The last Labour government sanctioned and even arranged mergers of big science-based companies on a level which in the 1950s would have been regarded as intolerably monopolistic. The argument is also used to support the UK entry to the EEC.

'What is misleading,' as Calder the scientific columnist points out,[45] 'is the idea implicit in Britain's European policy ... that political aims must bow to technological compulsions – in the creation of rival super-states, with built in pressures towards "big" technology.'

Sweden presents a shining example of how wrong such views are. A small nation of seven million people, much of it above the Arctic Circle, remains a leader in innovation as in social relations. Sweden has understood that the power to create wealth is the power to make decisions, and that innovation is the only way to exploit the social resources of the nation. They recognise that the moment all innovation is imported an industrially based community has handed its destiny over to others. Significantly it has not joined the EEC. Sweden, perhaps more easily than any other nation, could switch relatively painlessly from economic growth by volume to that through quality.

The concepts of economic growth have confused the relationship between identity, innovation, industry and independence. Nowhere is this more so than in Europe. While on the surface it is clean, bright, diverse, cultured and busy reorganising itself into a bureaucratic super-state, it is fast losing its grip on its psychological and social resources – the very things that Europeans do not wish to lose. Even Servan-Schreiber, whose famous *Le Défi Americain* first exposed and analysed the threat to Europe, has failed to see that the danger to Europe does not come from America, but from within itself. It is not America that is the enemy, but obsessive industrialisation. Servan-Schreiber argued that since, after fifteen years of the Common Market, the one country that had truly benefited was not any of the European ones, but the USA, here was a yet clearer demonstration that Europe must band together against America. He portrays in a vivid way the symptoms of cultural and economic collapse, symptoms far less advanced than those seen in present day Scotland. Let him speak for himself.[46]

> 'At times like these we naturally think about reinforcing the barricades to hold back the invader. But surely defensive measures might well make us even weaker. In trying to understand why this is so, we stumble on the key element. This war – and it is a war – is being fought not with dollars or oil or steel or even with modern machines. It is being fought with creative imagination and organisational talent.'

Yet his chauvinism is wrapped in a tinsel of materialism. 'If Europe continues to lag behind . . . she could cease to be included among the advanced areas of civilisation within a single generation.'

How is this consistent with the OECD forecast that France, his own country, would be the third richest in the world by 1980? Is he telling us that the GNP is not the sole criterion of wealth? We would agree.

> 'If we continue to allow the major decisions on industrial innovation or technological creativity – decisions which directly affect our lives – to be made in Washington, New York, Detroit, Seattle and Houston, there is real danger that Europe may forever be confined to second place.'

Servan-Schreiber sounds momentarily as plaintive as a spokesman for the Conservative party.

The French government has been no less greedy in seeking an expanding economy than has the British. Happily France is blessed with much better ratio of resources to people and has reasonably intact psychological resources. It is true that the French government once tried to curb American investment, but that was in the early days of de Gaulle. Here was a man who understood nationalism, understood the value of the psychological resources of the French nation, yet even he was forced to relent. He could not make France go it alone and expand industrially. It is important to see the connection. He was a victim of a pressure of big business and of the Trade Unions. Moreover, nationalisation, even for France, could be no solution.

'In so far as foreign investment merely reflects superior technology, we could nationalise only factory walls. You can't expropriate technical knowledge and inventive skills.'

'In the last analysis it is clear that the power to create wealth is in the power to make decisions . . . what we need to put us back into the race is organisational skill and a determination to be independent.'

'It may be that Europe lags so far behind in innovation and management that any attempt at economic independence would only retard us even more. But even if such a withdrawal from the fight were justified economically, we don't have to accept it. There are political, cultural and even moral arguments that should persuade us to reject this easy temptation to "Americanisation" . . . we see that independence is not an ethical notion, but an economic necessity.'

Servan-Schreiber has almost brought himself to utter the message of this book, but he returns abruptly to the strait jacket of economic expansion.

'In a modern world it is only in a large market that a sophisticated economy can be sustained.' When he says that we must have 'the courage to recognise that our political and mental constructs – our very culture – is being pushed back by this irresistible force,' he refers, we believe wrongly, to America and not industrialisation.

Servan-Schreiber envies the small, highly developed countries. 'The road taken by Sweden or Switzerland (of competition by specialisation) is not now open to France.' Does he perceive in them a balance lacking in the bigger industrialised states, or does he recognise their smaller size as more easily permitting a change of course? Nevertheless:

> 'Western Europeans look on self-determination as an acquired right; they cannot imagine it could really be threatened . . . the day this drive weakens to the point that Europeans let somebody bigger to their work for them, the spirit of our civilisation will have broken . . . we would be tainted with the knowledge of our own failure. Without suffering from poverty we would nevertheless soon submit to a fatalism and depression that would end in impotence and abdication.'

It is questionable, even if the concept of the expanding economy were to hold sway for many more years, whether the Common Market would be the answer to the risks that Servan-Schreiber sees to the social and psychological resources of the European nations. The USA will presumably continue to benefit so long as Europeans allow the industrial imperative free rein. In its fundamentally materialistic argument for the UK entry into the Common Market, the British European Movement points out that, of the EEC countries, Belgium before entry to the market had an annual rise of GNP as low as that of the UK but is today 'enjoying' one many times greater. It fails to record that Belgian industry is heavily penetrated by the Americans, that the standards of pollution control are the lowest in Europe, and that the country has effectively become an industrial colony and that riots have arisen as the Flemish-speaking population fight for the preservation of their psychological resources.

One wonders how eager are the Rotterdamers, a little to the north, to accept more industry at their already industrially overburdened mouth of the Rhine. Already by consensus they have rejected the proposal of an integrated steel mill.

'Science,' a past Minister of Technology told us,[47] 'must be harnessed to the job of earning our living as a nation.' Yet the argument on behalf of innovation has still a long way to go before it finds a place in political minds. Education is needed, and not merely amongst the politicians and Trade Unionists and money men.

Gunnar Myrdal, the Swedish economist, has pointed out[48] that an expanding economy on a technological plane, that is in terms of quality, will require massive re-training if there is not to be created a vast body of unskilled unemployed: an underclass. There are already signs of such a class emerging in industrial colonies such as Scotland, where it is commonplace for incoming industrialists to make remarks like, 'There are no unemployed here, only unemployables'. Education as an element in the survival of the nation is no longer argued, yet in quantitative terms it was first clearly identified, only as recently as 1964 by Edward Denison, when working with the National Council for Economic Development in the USA.[49] He saw that historically in the USA in the early part of this century, expansion was basically a question of numbers, and calculated that more than half the economic development between 1909 and 1929 was due to the expansion in numbers of people at work and the growth of invested capital. This was natural in a relatively empty, but well-endowed terrain. After the Depression this was no longer true. Between 1929 and 1957 these quantitative factors (labour and capital) were responsible for a mere third of the increase of GNP. Today the most important factors are education and technological innovation, with education the more important. Quantitatively Denison put education's contribution as 11 per cent at the out-

150 THE POLITICS OF ENVIRONMENT

set of the century, 23 per cent for the middle period to 1957, and much more now. It is therefore instructive to look at the figures of students in various countries, taken from the data compiled by Chorafas (*Brain Gain or Brain Drain*).

Table 8

EDUCATION

(from the Chorafas Report)

Country	Number of students in 1966	As percentage of population aged 20–24	As percentage of total population	Estimated order of wealth in 1965 GNP per head
United States	5,526,000	43	2·2	1
USSR	4,000,000	24	—	—
Japan	1,370,000	13·5	—	14
France	500,000	16	1·0	3
Italy	284,000	6·9	0·55	—
Germany	280,000	7·5	0·47	5
Canada	230,000	22·5	—	6
Sweden	62,000	11	0·8	4
Belgium	54,000	10	0·56	10
England and Wales	143,000	4·6	0·28	15
Scottish (Scottish origin)	22,000	7·0	0·43	20*
Britain as a whole	265,000	4·8	0·30	15

*There is a rough correlation between student population and wealth, except in the case of Scotland. The discrepancies may be partly explained by the technological strategies already adopted by each country and in Scotland's case by her effective colonial status.

Having posed several problems, let us in conclusion examine some consequences.

Had the world not become over-populated, the switch from growth by volume to growth by quality could well have been made slowly, intelligently and with minimum disruption. Such now seems beyond hope. Disruption for some countries and many economies seem inevitable.

There are, at the moment of writing, some 800,000 people out of work in the UK, more, as a proportion of the population, in Scotland than in the rest of the British Isles. Clearly the economy can continue without the aid of this idle work force. As industry and business evolves ways of being more and more efficient, an adequate output may be achieved with fewer and fewer people. If the economy were to stagnate or come into balance, even fewer would be needed.

This has created a situation in which business or industrial efficiency, far from being desirable, actually cuts the roots of survival. In all probability most people would rather be inefficient, but at work, than on the dole in an 'efficient' economy. Universities are very good examples of the way in which clever but inefficient people can be employed. Universities survive their monstrous inefficiency because they are not in competition in an economic sense. Their output is concerned with quality not quantity. Their reputation, and the ease with which they can attract good academics, depend on their high standards, not their index of production. By contrast, countries or groupings (such as EEC) are concerned with competition between each other, with grabbing trade, if necessary from someone else. In such circumstances it is extremely difficult for one country or group to withdraw unilaterally from volume expansion. However, eventually the shortage of world resources will reshape this situation. It will be a major tragedy for the world, perhaps even the final tragedy, if it does not of its own volition learn to control consumption and thus expansion. The problem of change is very much greater for highly developed countries, especially ones where the ratio of people to resources and land is high. For the less developed countries, the prospects are much better. They can learn from our industrial evolution. Here is an assessment of those nations which seem best equipped to survive.

The criteria are: balance of population and resources,

material, psychological and social. The following gives some useful comparative data.

1. *Those countries for whom there is the greatest hope are:*
   Finland, whose speed of achievement has probably no modern equal.
   Sweden, technologically very strong, but culturally threatened.
   Denmark, partly by virtue of its possession of Greenland.
   Norway, where use of even peripheral resources is actively encouraged.
   Iceland, who single-handed and with a population of less than Aberdeen, took on and won the battle for her marine resources.
   Austria, who has suffered enough to appreciate the consequences.

2. *Those countries salvable, but in immediate danger:*
   Canada, already culturally and economically invaded. However, she has such enormous resources and a growing national awareness of her problems and potential.
   Brazil: sufficient sense of nationalism to hold the invasion in check, seriously impeded by her present non-government.
   Czecho-Slovakia: a wonderful recovery, checked by Russian invasion.
   France: some use of nationality, offset by delusions of grandeur.
   Scotland: severely weakened and put off balance by long economic aggression from England. Culturally heavily penetrated. Community spirit still existing. However, it is the only part of the UK immediately capable of being saved.

3. *Those countries too late to salvage without suffering major disaster first.*
   England: increasingly committed to over-population, over-

industrialisation, racial problems and re-adjustment to eclipse of centuries of moral and physical superiority.

Germany: locked in forward gear. Very little control of the engine. Fortunately in the best seat.

Bene-Lux: cultural and community spirit cannot withstand the steam-rolling by the Treaty of Rome.

PART IV

3. Resource Management

Resource management is usually considered in the sense of material resources. It will be used here in the sense of all resources available to the community, physical, social and psychological.

It is probably easier for the physical scientist or technologist to visualise a physical limit to resources than for others. He is trained to think in terms of input and output, and if his education has been at all successful, he will understand the import of the Second Law of Thermo-dynamics. Thus, no matter how ignorant he may be of biology, he can comprehend through this Law, the argument that the species highest on the life pyramid are the most susceptible to shortage of energy and hence in the most precarious state. It is to be hoped that the economists who devise means of manipulating resources will find means of arriving at the same innate understanding. Their present dominant status as advisers make this imperative.

For the science community at large it is still a comparative novelty to be discussing and pontificating upon how the world should organise itself. This is something it has been content to leave to others, and the others have seldom been pleased when the scientist has come out of his laboratory and started to take part in the discussions. Only recently a scientist, appearing on British television, remarked that the royal family, by having four children, had failed to set an example in population stability through self-control. He found himself the subject of a leader in the *Daily Express*, telling him to go back to his bench and leave wider issues alone. Fortunately, the science community is not doing this, and as it grapples with the problems its inventions have created and other men have so misused, it becomes more articulate. Through the menace of industrialisation, science and morality are face to face. An influential cross-section of them met at Aspen in 1970, and after many days of deliberation unleashed upon the waiting world the following :[50]

'The rapid growth of population and its concentrations in some parts of the world are particular and growing causes for concern. There is an urgent need to move as quickly as possible towards a stationary population, not merely for the improvement of the quality of life but for the development of human personality. Only by the control of population growth can societies hope to share the potential benefits of technology. Moreover, we know that economic and social progress is one of the most effective inducements to a declining birth rate.

'We believe it to be important that the encouragement of economic growth should not be simply an end in itself, but also an encouragement of social development.'

The problem they did not face up to was the time lag between knowledge and application, between the use of technology and its long-term effect. And they did not face up to the means.

Control of resources
For as long as one cares to think back, wealth has lain in a command of resources. Latterly this command has not been so much in military power, but in the power of finance. Canada's enormous store of resources has brought less riches to her than to those who have financed their exploitation. But recent events in the oil-producing countries have shown that, if a resource becomes too widely used, if it becomes too essential to an importing economy, and if it becomes too limited in amount, then the supplier of the resources may begin to regain power. Military action as a means of economic expansion died a natural death with the British-French-Egyptian war in Suez. Today, the presence of nuclear threats reduce war to a

mutually destructive activity. It could still occur, but no longer for economic reasons on the part of the developed countries; they have too much to lose.

Power will move from the consumer to the producer. The developed countries, especially those of them who are massive importers, will need to re-think their situation. It is in their interest to seek world resource control while they still retain some measure of control through financial power and consumption. It will not be easy, for control on consumption may be necessary and such control implies a limit to economic expansion. The international agreement on whale hunting may serve as an example of the problem. While biologists set the hunting limits of the whale to a number which would sustain the species, the governments of the participating nations were successfully pressed by their respective industries to permit greater numbers of killings, even to several times the biologically sustainable yield. Thus the world whale population has fallen dramatically in the last few years. Only Russia and Japan are left in the business. Ehrlich comments, 'Their drive towards self-destruction tends to contradict the commonly held notion that people would change their behaviour if they realised that it was against their own interest'.[51]

Something similar is happening to the herring industry in the Atlantic. It is an irony that our inshore fishermen, who have collectively set out to conserve inshore waters, and who have won, through years of Scottish lobbying, a twelve-mile fishing limit round their coast, are in danger of having their self-control wasted as these waters are thrown open to Common Market countries.

Though control of resources may be achieved in the end, human behaviour hardly gives us confidence to think it will occur soon enough, nor without more frightful waste. In the short term, and that may mean several generations, those countries that have a reasonable balance of resources to population stand the best chances of a decent life. It is interesting to see how relative situations can change with the advent of new technology. Again, taking Scotland as an example: for the greater part of this century Scotland has been run down industrially. It was a situation regarded by many as inevitable. Even in the 1960s many Scotsmen would sorrowfully observe that their country was no more than a peaty hiccough on the edge of Europe. To them it made no economic sense. How speedily matters have changed in a decade! Today, the oil being produced from the Scottish sector of the North Sea exceeds Scotland's needs, making her one of the few European nations, had she her own government, independent of imported oil. The Canadian arctic, one of the great reserves of mineral resources, is as convenient to Glasgow as to Montreal. In an era when giant cargo ships are being built to make cheaper the movement of raw materials, problems of collision and its shallow water renders the English Channel impassable, or at best dangerous. Europe finds its best port lies on the west coast of Scotland, a place where even one million ton ships can berth in sheltered water only a short distance from the shore. Water, a rapidly waning resource in much of Europe and England, is still abundant in Scotland. Pollution outside the midland belt is negligible.

Control of pollution
In 1970 opinion polls in the USA showed that 56 per cent of Americans favoured spending more on pollution control, putting it next to crime prevention in their priorities. It would be interesting to conduct the same enquiry, phrasing the question differently; to ask if they favoured pollution control by cutting back industrial production, that is, creating a self-imposed recession. It is very doubtful if the answer would be anything like so positive. Positive answers would come from the foresters in Scandinavia, whose forests suffer from the pollution of UK factories; from the tuna fishers on the west coast of the US; from the salmon fisheries in England, or from the Bavarians polluted by the Ruhr.

For the moment such pollution control as exists is very much a result of cost benefit analysis and tends to be a direct result of organised protest. Technology is the scapegoat. The real culprits slip away and the environment becomes the great talking point of the age for want of anything deeper. Yet even here only a handful of people are talking reality, the rest are still at the stage of mouthing platitudes. For the environment is something people seem to sense is bound to be sold out. The action of the conservationist is not so much conservation *in toto*, as a rearguard action to slow down destruction. 'All right, you can have half the forest this year, if you leave the other half for another ten.' The urban dweller who needs and loves his countryside is still convinced that it must go in the lava tide of industry, which he has been conditioned to believe is the first priority. True, concern for the environment is manifested at government level in most developed countries. The arguments become more sophisticated upon both sides. There are remarkably few examples of protest effecting any more than an official 'undertaking' to keep pollution at a minimum. Placating the public is now a matter of good public relations before the job is done. The threatened community is easy to deal with, as the central government found over the proposals for industrial development at Hunterston in Western Scotland, locus of fabulous deep water berths. There was high unemployment in the area. Anyone who protested was condemned as being against giving people work. The fact that the new developments projected there would bring work largely to incoming Englishmen was never mentioned. The 'unemployables' would thus remain. In an intellectually developed community, not so threatened, the result is different.

In 1968 a US power company proposed to use deep Cayuga Lake (New York State) for cooling water. Aware of the need to placate local opinion, it indulged in a careful publicity programme designed to show no serious pollution would occur, and no other harmful effects would arise. It employed 'experts' to back up these opinions, and just to emphasise the value of the power station, pointed out that its presence would hold down power costs to the consumers in the immediate area. Fortunately the calibre of people using and living by Cayuga Lake enabled enough informed counter-opinion to be brought in to modify the scheme.

Informed attacks such as these bring out how valuable land, air and water are to any community. Commenting on the Cayuga affair, Alfred Eipper says:[52]

> 'These decisions must be basically public decisions, and cannot be made unilaterally by any particular interest group, be it industrialist or preservationist. With increasing population and declining resources, Resource Management must become a principal function of government and of the UN. This may be the most vital purpose of government in modern times. When considering the development of any activity which involves the environment, and almost everything from a farmer's fertiliser to the creation of an aluminium smelter, does so, the narrow conventional economic criteria are at best inadequate, at worst disastrously misleading if used as the sole basis for decisions.'

And, as stated by a committee of the US Congress:

'The market approach fails for two reasons: first it is very difficult to quantify in dollar terms many of the values of environmental quality. Second, the axiom that a unit of profit is more valuable now than at any time in the future leads to short-sightedness in environmental management.'[53]

Many US states now forbid the processing of crude oils that contain more than one per cent sulphur without special provision. The result is New Englanders find it hard to get fuel oil for their central heating without importing it from abroad, and US oil companies are looking for sites in any country desperate enough to lure industry and keep mum about pollution. They exploit the 'ethical' gap, as Nigel Calder so nicely puts it, between the standards of the technologist on the one hand and the so-called 'realistic' standards of social and political life in an unbalanced community. They exploit the fact that work is wanted, that people crave jobs which will give security, higher incomes, and that most politicians think only as far as the next election. It is practically impossible to oppose them, because both the politicians, left or right, and big business have a common interest in selling production. Production is vote-catching.

What is important here is the widespread belief amongst a certain portion of the community that it does not matter too much about pollution because science and technology can solve *any* problem; that it is just a matter of time, money and effort. This blanket appraisal, which finds little support amongst the scientists and technologists themselves, is convenient to politicians. It is as if by the penance of a few hundred millions of pounds, any of mankind's excesses can be comfortably expunged. When the rich ores are gone, we can use the lean ones. If a smelter destroys the trees, then research will provide a wood substitute. It is ludicrous to desire both colour television and freedom; why not settle for the former, which can be distributed to all?

But technology is creating such powerful instruments that nations must stop being parochial. If Florida uses silver iodide to create rain, will that create a desert elsewhere? Who created the mercury pollution ingested by the tuna in the Gulf of Alaska? If supersonic jets are found to put enough carbon dioxide into the tropopause to alter the world's heat balance, all the world is going to object, but it may be too late. Since the world is overcrowded already is it wise to breed freely as we do? Will China accept one standard while Brazil demands another? The oceans are common property. Can every country just pollute them freely? Is it dangerous to allow computer networks to be built? They may become a tyranny we cannot be rid of.

Control of resources
The remedy is not easy, nor does it consist simply of the con-

trol of pollution. None of us will willingly take a step down in our material standard of living, though we might accept it communally if a good case were presented to us, showing counteracting benefits or if neighbouring nations were in the same fix. Surely, as Calder says,

> 'Growing wealth and technological powers should multiply, not diminish personal choices. A moral and political test for the new engineering will be whether it aligns itself with the "system" of god-like social planning or with the individual, his idiosyncracies and his psychological quirks and fancies. . . . Technology may have an inbuilt tendency to totalitarianism, but we cannot seriously contemplate doing without technology. Moral and political efforts need therefore to be directed to compensating the totalitarian tendency, and to employing governments to encourage human brands of technology.'

One's freedom is in any case limited by the boundaries of one's fellow men. Does it make sense to restrict it further by letting the expansion of industry take over from us the right to use new technological developments as we, our community, see fit. A 'united Europe' will, as its protagonists allege, make wages rise, industry grow and increase the movement of people. They do not tell us if we shall be richer and freer or have a better environment, or run less risk of war, or enjoy better health.

Freedom is the converse of efficiency. Efficiency here is used in the true sense of the word; that is, what is obtained or achieved divided by what could be achieved from an ideal system in which extraneous interference is absent. No system has been found that is a hundred per cent efficient. Efficiency as the word is usually used, is in reality effectiveness. All electric heaters are practically one hundred per cent efficient in terms of converting electricity to heat, but their effectiveness varies greatly. An efficient system dominates the user. An efficient bus service would run with every seat occupied. An efficient farm obtains the highest possible yield from the given soil. An efficient business maximises profit per unit of capital used. In none of these examples is the efficiency measured in terms of the individual. It is measured in terms of inanimates or abstractions; the bus company, the soil, the firm. In achieving efficiency for these abstractions what has happened to the effectiveness of the human being?

It would be a comparatively simple matter to arrange by international agreement to get a sustained harvest from the oceans by quasi-industrial methods, using comparatively few men and a lot of machinery. If it were to be done, some people anyway would go around proclaiming that the fishing industry had been made more efficient. The fact that many fishermen would be redundant would not have entered into their calculation of efficiency. However if as a result of such industrialised fishing, more fish could be harvested on a sustained basis for ever, it is certainly arguable that since the world is short of protein such fishing is permissible or even desirable. But such is not the situation. The world's ocean harvest is possibly already too great to be sustained. Pollution is not easing the problem. Is it not better to leave the fishermen to their inefficient methods? There is little alternative activity for them and certainly none which breeds people of such vigour and quality. As a genetic stock they contribute a great deal more to the community than the fish they catch. By permitting them to fish by 'inefficient' methods, they remain men, effective men. But out of work they wither inside. Put them in factories and a similar change would occur. But since the economy cannot expand, where does one find the factories?

Similar arguments can be used for the preservation of marginal farming in hill country. Moreover, the activities of these farmers brings a benefit, never brought into the calculations. They nourish and crop the hillsides, creating pleasing patterns and an environment with which to delight the city dweller on his holidays and weekends. It costs no more to maintain him in his inefficiency than to meet his weekly unemployment cheque.

The Mansholt plan for European agriculture cuts right through this argument, for it bases itself not on the effectiveness of the individual but on the efficiency of the economy, an abstraction that has no justification in human terms.

The psychological resources have been examined in an earlier part of the book. The need for their preservation should surely not require further argument. We suggest two events to give food for thought. They both occurred in thinly populated areas.

In 1903, Lord Leverhulme, owner of much of the island of Harris, decided he would raise the people out of their uneventful life by means of industry, and caused a margarine factory to be built at Obe. He had the bad taste to rename the township Leverburgh, but to give him his due, his intentions were of the best. The project died because the people would not work there. They foresaw that, despite apparent affluence, they would have sold their community to Lord Leverhulme who owned the land. Had they been able to enjoy control over the community they so deeply loved, and which drew from such rich and stable cultural roots, Harris might have assimilated Leverhulme. But, wisely, they called off and Harris, neglected and not yet ripe for 'development', is still, but only just, alive.

In 1971, Alaska, with half a person to the square mile, is fighting tooth and nail an oil consortium which proposes to build a 700-mile pipe-line through tundra country from the North Slope wells to Valdez on the Pacific. Why are they fighting it? There is surely space for everyone including the indigenous Aleuts. It is because people who go to Alaska are all highly motivated. Either they go there to make money and want or are indifferent to the pipe-line; or they want to build something better than the mess that exists in the rest of the USA and care deeply about their surroundings.

PART IV

4. The New Politics

Peter Ustinov, addressing the students of Dundee University, after they had elected him their Rector in 1968, declared that the future of the world lay in the hands of the small nations. To many of his listeners it seemed as if nothing could be farther from the case.

Now it seems like a revealed truth. Large economic groupings have no justification if their concern is only for themselves. International control of resources and pollution seem inevitable. The need for innovation and diversity points to smaller units. World government is doubtless a desirable objective, but it can never be one single, central government dominating every citizen. Such would be truly unmanageable, and the consequences if power ever fell into the hands of an oligarchy are too dreadful to contemplate. Stability lies in diversity, each unit inter-related, inter-dependent. Barnaby Keeny, chairman of the National Endowment of Humanities in Washington, commenting on the fact that it takes a generation or more for new interpretations to find their way into school books, complained that :

> 'Historical knowledge is not used by today's policy makers. It has been known, for example, for at least a generation, that representative democracy had its origin in local government and that the key event was the discovery of a way to represent that local government, which had a strong popular base, in a representative assembly.'[54]

In the name of efficiency local government is being destroyed all over the world, even in supposedly freedom-loving and democratic England. The journalists of the Aegnor reporting team in Brussels express beautifully the confusion between efficient and effective.[55]

'Self management is a basic political option. It springs from the belief that new forms of democracy are needed, adapting to the scale and complexity of the modern world the principle that people should decide for themselves. . . . Except in a totally closed economy it is in the workers' interest that self-management should ensure at least as high a level of economic efficiency as competing systems. Only an economically efficient system will be politically free to make choices about the use of its resources for anything other than ever-increasing wealth.'

Workable democracy is needed more than ever to see that technology and industry operate for the benefit of the people. The elected representative must be educated and competent. Of the UK parliamentarians, Calder bluntly remarks:

'. . . a member who is a qualified scientist is a curiosity. The typical parliamentarian is elderly, arts-trained, unaccustomed to sustaining deep thought, and equipped with political rules of thumb scarcely more relevant to shaping affairs today than the theory of phlogiston is to modern chemistry.'

Some members today may still be literate, but few are numerate.

No matter where the discussion starts we are drawn to political solutions. Every country must upgrade the quality of its parliamentary representatives, for they directly reflect the standard of government. Moreover, to quote Keeny:[54]

'There is a need for new political instruments. Our present political instruments may be adequate to adapt to (this) change, but I see as yet no evidence that they are, for one element that is utterly lacking is trust – trust between individuals and trust between societies that makes negotiation and compromise possible. It is also possible to denounce the suggestion of a need for new political intsruments as subversive, and I have no doubts that they will be so denounced, but that does not alter the facts of the situation. How are we to cope otherwise with the social results of medical progress and to control the unceasing growth of population... ?'

The problem of creating a change in the political order is great, and even more so in a democracy. We have first to bring about a fundamental change in values. We must combat, as Professor Chomsky puts it,[56] 'the psychotic world view that has been constructed to rationalise the race into destruction'. Traditionally it is the Left that has tried to do things about this;

'but in the long run, a movement of the Left has no chance of success, and deserves none, unless it develops an understanding of contemporary society and a vision of a future social order. . . . In an advanced industrial society it is far from true that the masses have nothing to lose but their chains . . . they have a considerable stake in preserving the existing social order. Correspondingly the cultural and intellectual level of any serious radical movement will have to be far higher than in the past.'

The business leaders, the sociologists, the economists, the politicians are all at odds in their attempts to deal with and explain what is happening. Goebbels recorded in his diary that, 'The fight against unification of Europe by the Axis powers is mainly carried out by the intellectuals. The broad mass of people are uninterested.' It remains true. The intellectual's fear of the future is openly voiced, though not yet fully expressed. It is still,

'a brave public figure who is opposed to technical progress in general. . . . (Yet) in the face of such dismay, the consensus cannot survive, but it is unlikely that the old political ideologies will prove to be satisfactory vehicles for the controversy. On the Right, traditional conservative ideas cannot be sustained now that so much research and technology for industry requires state support, now that in a world committed to change the stability that has to be conserved is more like the stability of a jet aircraft than the stability of a castle. Traditional parties of the Left are particularly vulnerable to both the consensus and to the direct effects of technological change. The first is plainly destructive of the parties' ideology. The other erodes their sources of power among industrial workers by absorbing skilled workers into near-managerial status in science-based industries or automated plants and threatening others with large scale redundancy'.[57]

In searching for the answers it is encouraging to find that at least the scientist and the humanist have found the same wave-length. Here is McElroy, the director of the National Science Foundation, the world's largest science funding agency:[58]

'I sometimes wonder about the dangers of those who exult reason at the expense of all else, with those who analyse a rainbow, and miss the magic of the colour. In my view the science community generally should consider carefully . . . (what has been termed) . . . the new romanticism, emphasising man as an emotional and feeling creature as well as a reasoning one . . . (though) disciplined rationality is at the heart of man's best work in science, even our scientific radicals overturned tradition through the process of painstaking reason. From the vantage point of history, scientific reasoning coupled with creative imagination is a powerful tool.'

Looking round, one finds no political party in the United Kingdom, nor in other countries, whose policies appear to have the slightest relevance to the current circumstance of man. Mr Rippon, the Conservative government's Common Market negotiator, woos the electorate by telling them that in the EEC Britain will once again take 'her rightful place among the foremost nations of the world'. The world is too small and sophisticated for such jingoism. The Conservative Party must founder upon this basic tenet of free competition. Though the serene confidence of 1900 has worn off, the theories of Adam Smith and Darwin still provide for this party the old fashioned but convincing rationale for 'organisations, values and operations of the bourgeois and imperialist systems'. It is the party that supports capitalism, that leaves men and industry in a mutually destructive scramble; de'il tak the hindmost. So far as the circumstances will allow, its supporters like to believe they are permitting the market system to operate. The party regards itself as the defender of tradition, while their offspring throw ink over every tradition within sight.

Free enterprise is a superb incentive to the industrious and ambitious man. It elevates his whole capacity as an individual. But it is necessarily done at the expense of his environment, human and physical. When the world was a large empty place, this was to some extent permissible and very agreeable for the individual. Social instruments were anyway not developed to oppose him. In South and Central America, the Spaniards, armed with no more than a Catholic creed and a Papal requisition, took over the better part of a continent for their enrichment. The Protestant ethic made a similar successful depredation on North America. In an empty land, the first settler would appear to have some 'right' to the resources of the territory, to use, despoil and reject as he wished. Thus did he create the dust-bowl of the Mid-West of America. But the moment that his actions impinge on others, then the arguments for monopoly depredation fail. In any crowded country, as in Europe, every act affects one's neighbours. For example, after ten years of legislation it is quietly accepted that it is anti-social to burn a coal fire in a city grate. It pollutes the air breathed not only by the firemaker but also by his neighbours.

There is nothing in the Conservative Party philosophy that prepares it for controlling technology for the good of all. The performance of the Conservatives since coming to power in 1970 has demonstrated this well. They regard technology, not as opportunity for investment, but as a sink for wealth. Their cut back of support to scientific and development institutions, especially at a time when the United Kingdom was facing a serious economic crisis, will be shown by history to be the beginning of their ultimate eclipse. Finally, it is the party which is supported by big business, which provides the bulk of its funds. It is unlikely that the Party will bite the hand that feeds it, and yet it is up to a political party to lead the nation in new thought. By its very nature and support, the Conservative Party cannot convert from economic growth by volume to economic growth by quality, though a growing number of its civilised supporters will want to do just that and many are ardent conservationists. They cannot have their cake and eat it.

By contrast, Labour reverenced Big Science. Being a party committed to planning the economy, it did so in detail. It saw more clearly than its predecessors the form that cultural and economic aggression would need to take if Scotland's resources were to be quickly available to England, and is more responsible than any other party for the present appalling subservience of the Scottish economy to external influences. And yet, for the Labour Party, technology is the greatest problem. Willard Wirtz has remarked that for the human being to compete with the machine, he would need at least a school-leaving certificate. We see looming before us slag heaps of unqualified people. They form no sound basis for a progressive political party. Since Labour claims to speak for the worker, William Wolfe is entitled to his remark that, 'The party that puts so many people on an emigrant ship cannot be truly called the party of the working class'. The Labour Party has traded the concepts of human equality for those of industrial production. It buys its votes, as surely as do the Conservatives, by committing itself to an unthinking and continuing economic expansion. In power it did nothing to combat the irrationality of the free-for-all abuse of contemporary technology, and the enormous bias of the market system against the individual. Labour contributed to the contemporary barbarism, so condescendingly expressed by a Conservative Prime Minister in the words, 'You've never had it so good'.

The Old Left has let down its supporters by not only failing to be radical any more, but in abandoning morality. Where are the high principles of the pre-war international socialists, willing to band together for the good of all? Sir Leslie Cannon, a few days before his death, remarked sadly that 'The Trade Union movement has lost its way . . . its sense of mission . . . the powerful get what they can and leave the weak. . . .'[59] The Party morality is at its lowest in Scotland, where it is tacitly accepted that the best way for Scotland is to twist Westminster's arm to obtain the maximum charity, heedless of the illogicality of this attitude. As Scottish Trade Unions submerge in the flood of English mergers, their ambitious leaders keep their eye squarely on the top jobs in England. Even a jackal's share of the power. . . .

It is doubtful if there is a truly radical party in the British Isles, putting aside the Anti-Imperialist Solidarity Front, the Trotskyites, the Workers' Party of Scotland and numerous anarchistic organisations. Their radicalism lies only in the fervour of their pamphlets and the upheaval their 'solutions' would occasion. They are neither better nor worse than the principal London based parties available for re-election: probably more interesting, but equally irrelevant to the future. As radical parties aver, all men are equal. But it is pure romanticism to imagine that one can create a vigorous equitable society through class warfare, social strife and a pre-occupation with wage bribery. Man needs an ideal by which to live and moral leadership. For many years now the Labour Party has failed to give either.

Without conflict and ideals man withers. He is reduced to manipulating the system. The unskilled workman unfortunately seems no more immune from the basic instinct to work it to his best advantage than the stockbroker, estate agent or professor. In our socialised society people will devote real energy to design a situation that will maximise their own social security benefit at the expense of their self-respect.

Planning, of course, is needed, provided its purpose is to benefit the community as a whole. The Labour Party has used it as an excuse for unwarranted centralism, led not by Parliament, but by a powerful bureaucracy.

As science becomes more complex, and politicians become less competent to cope with it, the inevitable trend of centralist government is to become patronising, at best avuncular, at worst dictatorial. We live in a frightening time for democrats. Too many people appear to be willing to trade democratic rights for hand-outs. One hears too often that 'the government ought to be run by a team of experts', or that 'we need a more authoritarian government to deal with strikes and vandals and crime'. These are opiates not cures. At the last Westminster election the people voted for a change of government as much on this premise as on any other.

The Conservative and Labour parties are seen more and more as two dinosaurs ponderously pushing each other out of one era into a newer one in which neither is relevant. The task facing the living communities in the British Isles is to prevent their last blind convulsions from irreparably damaging our heritage. Technology is radical, and technology is expanding. It is this we have to learn to use and control, and to control democratically. New parties are needed which can provide a more rational relationship between men and their environment. Calder sees the realignment in horizontal terms as between Technological Opportunists and Scientific Conservationists. The former are dedicated to an imaginative leap into the future on the basis of undreamt of technological progress, a world in which no problem created by man cannot be solved by sufficient time and effort. The latter are committed to the preservation of human and environmental well-being before material progress.

The speed of re-grouping round the new ideologies will depend on the educational level of the community and the degree of imbalance. Change will occur more slowly in imbalanced communities because of the vast upheaval a change from the present will create. The wood will be screened by the trees falling in the economic gale. Such communities will need the example of more enlightened communities to show the way. It will have a greater chance amongst the intrinsically balanced communities, such as Scotland, where there already exist political re-groupings. Scotland, always a radical country at heart, may see the issue at its face value and usher out the dinosaurs. It already has a radical, widely-based and growing political action group – the Scottish National Party – which can give people in Scotland a lead in the race for survival – if they use it. But the SNP, though it correctly perceives the need for self-government in Scotland, does not always understand the true reasons as several of its policy documents reveal. But these are minor drawbacks. Such policy documents are simply a reaction to the game of party politics through policies that Labour and Tory play at; a trap into which an inexperienced political party easily falls.

Action lies initially with the leaders of the community and we address ourselves to them. Upon the Trade Unionist leaders there lies a special charge. Their power over an uncritical and easily led mass of voters, so amply demonstrated in many a senseless strike, makes them key figures in the great re-think. For the moment the blind lead the blind. It is our hope that from an understanding of the problems and facts exposed in this book they will see that in the future world the preservation of each national community is vital to the well-being of their work people.

Without a sense of community there can be no evolution; it becomes merely a rearrangement into class factions whose leaders employ currently acceptable jargon for reviving the old aggressions and imperialism.

Such retrogressive behaviour as the world approaches saturation will assuredly destroy our species.

The true international socialism is of interdependent communities respecting their own parcel of environment, physical and psychological, and so ensuring for each of their men and women a comprehensible background against which to work out his or her own development. These communities will tackle their own internal problems and as they all share the same global environment, will be forced to share the responsibility for maintaining it.

Survival in the new world is clearly demanding new political and economic systems and these can only be practically developed in small, balanced, modern communities of the sort we have been describing. These communities must produce – and rapidly – the men to lead us to that goal which is so much nearer than the Moon and so very much more important, the World State.

We have seen that Scotland can be among these communities. Scotland, whose love of democracy and of freedom has so wonderfully survived the centuries, is well fitted for the future.

Scotland's environment is forcing her to develop her politics. The politics of environment go very much farther back than Marx or Wallace, and go very much farther ahead.

REFERENCES:

1. D. N. McVean and J. D. Lockie, *Ecology and Land Use in Upland Scotland*, p. 3, Edinburgh University Press, 1969.
2. Nigel Calder, *Technopolis*, p. 143, Panther, London.
3. A. Spilbaus, 'Man and Management in the Computer Age'. *Conference Board Record*, Feb. 1970, New York.
4. Hasan Ozbekhan, System Development Corp., California, USA, 1965.
5. E. Jantsch, 'Symposium on Technological Forecasting'. Tripartite Chemical Engineering Conference, Montreal, Sept., 1968.
6. Iain Henderson, *Power without Glory*, p. 83, Hutchinson, London, 1967.
7. J. P. McIntosh, *The Devolution of Power*, Penguin, London.
8. Scottish Development Department. Report of working party on migration in Scotland.
9. Cmnd 2188, HMSO.
10. John Butt, *The Times*, 11.2.1969.
11. *Economist*, 14.10.1968.
12. R. H. S. Robertson, *Eugenics Review*, July 1960. No. 52, p. 82.
13. J. P. McIntosh, *Scotsman*, 20.11.1956.
14. R. H. S. Robertson, *Eugenics Review*, July 1960, No. 52, pp. 71-83.
15. Kenneth Clark, *Civilisation*, p. 258, John Murray, London, 1969.
16. *Eugenics Review*, ibid.
17. Harold Lind, *Scotland*, Oct. 1969, p. 52.
18. *Tayside; Potential for Development*, Scottish Development Dept., HMSO, 1970.
19. Scotland's Government, Scottish Constitutional Committee (Conservative Party), Edinburgh, 1970.
20. G. McRone, *Scotland's Future*, Blackwell, Oxford, 1969.
21. (Foreword) Highlands and Islands Development Board brochure, 1970.
22. *Scotsman*, 1.12.1968.
23. *A Scottish Budget*, HM Treasury, 1969.
24. D. R. Donald, *Scotsman*, 15.8.1961.
25. Parliamentary answer.
26. *Scotland*, October 1970, p. 35.
27. E. J. Kermondy, *Concepts of Ecology*, Prentice-Hall, New Jersey, 1969.
28. *Resources and Man*, National Academy of Sciences, W. H. Freeman, San Francisco, 1969.
29. G. Borgstrom, *Too Many*, Macmillan, New York, 1967.

30. P. R. and A. H. Ehrlich, *Population Resources and Environment*, W. H. Freeman, San Francisco, 1971.
31. *ibid*.
32. J. Platt, *Science*, No. 166, p. 1115, 1969.
33. *Resources and Man*, p. 130.
34. *Economist*, Nov. 1970.
35. G. Borgstrom, *ibid*.
36. S. Enke, *Science*, No. 164, p. 798, 1969.
37. H. J. Barnett and C. Morse, *Scarcity and Growth*, John Hopkins, Baltimore, 1963.
38. W. Rostow, *The Stages of Economic Growth*, Cambridge University Press, London, 1960.
39. K. E. Boulding, *Environmental Quality in a Growing Economy*, Resources for the Future, Washington, DC, 1966.
40. A. Wagar, *Science*, 5.6.1970.
41. Henry Ford, *New York Times*, 3.12.1969.
42. E. J. Mishan, *The Costs of Economic Growth*, p. 62, Pelican, London, 1969.
43. L. Dice, *Science*, No. 170, 20.11.1970.
44. J. J. Servan-Schreiber, *The American Challenge*, p. 69, Pelican, London, 1969.
45. N. Calder, *Technopolis*, p. 48.
46. J. J. Servan-Schreiber, *The American Challenge*, p. 13.
47. A. Wedgwood-Benn, *Hansard*.
48. G. Myrdal, *Beyond the Welfare State*, Duckworth, London, 1958.
49. E. F. Denison, Brookings Institution, 1967.
50. Aspen Conference Statement, 1970.
51. P. R. and A. H. Ehrlich, *ibid*.
52. A. W. Eipper, *Science*, No. 169, p. 11, 1970.
53. 'Managing the Environment', 90th Congress, 2nd Session, 1969, USA.
54. Barnaby C. Keeney, *Science*, No. 169, p. 26, 1970.
55. 'Industrial Democracy', Aegnor Team, Brussels, 1969.
56. Noam Chomsky, *Liberation*, No. 14, pp. 39-43, New York, 1969.
57. N. Calder, *Technopolis*.
58. W. D. McElroy, 'Responsibilities of Reason', Indiana University. (Recorded in *Science*, No. 169, 30.10.1970.)
59. Leslie Cannon, *The Times*, 8.1.1971.

For Product Safety Concerns and Information please contact our EU
representative GPSR@taylorandfrancis.com
Taylor & Francis Verlag GmbH, Kaufingerstraße 24, 80331 München, Germany

www.ingramcontent.com/pod-product-compliance
Ingram Content Group UK Ltd.
Pitfield, Milton Keynes, MK11 3LW, UK
UKHW021438080625
459435UK00011B/302